Understanding Hydraulics

Understanding Hydraulics

Contributors

Yongfu Xu et al.

AURIS Reference

www.aurisreference.com

Understanding Hydraulics

Contributors: Yongfu Xu et al.

Published by Auris Reference Limited

www.aurisreference.com

United Kingdom

Understanding Hydraulics

ISBN: 978-1-78154-908-7

British Library Cataloguing in Publication Data
A CIP record for this book is available from the British Library

Printed in the United Kingdom

Exclusively distributed by CBS Publishers & Distributors Pvt. Ltd.

Sales & Distribution Rights only for India, Pakistan, Bangladesh, Sri Lanka, Nepal and Bhutan.This book is not to be sold outside these territories.

Contents

List of Abbreviations .... *vii*

List of Contributors.. *ix*

Preface... *xi*

Chapter 1 **Unsaturated Hydraulic Conductivity of Fractal-Textured Soils**............. 1

Yongfu Xu

Chapter 2 **The Impact of Ice Cover and Sediment Nonuniformity on Erosion around Hydraulic Structures**... 69

Peng Wu, Jueyi Sui and Ram Balachandar

Chapter 3 **Role of Hydraulic Conductivity Uncertainties in Modeling Water Flow through Forest Watersheds** 93

Marie-France Jutras and Paul A. Arp

Chapter 4 **Five Things You Didn't Want to Know about Hydraulic Fractures**..... 115

Vincent M. C.

Chapter 5 **Hydraulic and Sleeve Fracturing Laboratory Experiments on 6 Rock Types** 129

Sebastian Brenne, Michael Molenda, Ferdinand Stöckhert and Michael Alber

Chapter 6 **Fractures and Fracturing: Hydraulic Fracturing in Jointed Rock**....... 141

Charles Fairhurst

Chapter 7 **Estimating Hydraulic Conductivity of Highly Disturbed Clastic Rocks in Taiwan**... 179

Cheng-Yu Ku and Shih-Meng Hsu

Chapter 8 **General Hydraulic Geometry**.. 199

Levent Yilmaz

Chapter 9 **Hydraulic Fracturing in Formations with Permeable Natural Fractures** 213

Olga Kresse and Xiaowei Weng

Chapter 10   **Hydraulic Conductivity of Semi-Quasi Stable Soils: Effects of Particulate Mobility** 237

Oagile Dikinya

Chapter 11   **Electrokinetic Techniques for the Determination of Hydraulic Conductivity** 251

Laurence Jouniaux

Citations 279

Index 281

List of Abbreviations

ADV	Acoustic Doppler Velocimetry
AE	Acoustic Emission
DFN	Discrete Fracture Network
EGS	Enhanced Geothermal Energy
ESP	Exchangeable Sodium Percentage
GCD	Gouge Content Designation
HC	Hydraulic Condtivity
HF	Hydraulic Fracturing
LEFM	Linearly Elastic Fracture Mechanics
LPI	Lithology Permeability Index
MPM	Meyer-Peter and Muller
NF	Natural Fracture
OM	Organic Matter
PFC	Particle Flow Code
PSD	Pore-Size Distribution
RHC	Relative Hydraulic Conductivity
RQD	Rock Quality Designation
SAR	Sodium Adsorption Ratio
SNR	Signal-To-Noise Ratio
SWCC	Soil-Water Characteristic Curve
SWD	Soil-Water Diffusivity
TKE	Turbulent Kinetic Energy
UFM	Unconventional Fracture Model

List of Contributors

Yongfu Xu
Department of Civil Engineering, Shanghai Jiao Tong University, Shanghai, China

Peng Wu
Environmental Systems Engineering, University of Regina, Canada

Jueyi Sui
Environmental Engineering Program, University of Northern British Columbia, Canada

Ram Balachandar
Civil and Environmental Engineering, University of Windsor, Canada

Marie-France Jutras
Faculty of Forestry and Environmental Management, University of New Brunswick, Fredericton, Canada

Paul A. Arp
Faculty of Forestry and Environmental Management, University of New Brunswick, Fredericton, Canada

Vincent M. C
Fracwell Llc, Golden, Colorado, USA

Sebastian Brenne
Ruhr-University Bochum, Germany

Michael Molenda
Ruhr-University Bochum, Germany

Ferdinand Stöckhert
Ruhr-University Bochum, Germany

Michael Alber
Ruhr-University Bochum, Germany

Charles Fairhurst
Senior Consultant, Itasca Consulting Group, Inc, Minneapolis Minnesota, USA
Professor Emeritus, University of Minnesota, Minneapolis, Minnesota, USA

Cheng-Yu Ku
National Taiwan Ocean University

Shih-Meng Hsu
Sinotech Engineering Consultants, Inc Taiwan

Levent Yilmaz
Technical University of Istanbul, Civil Engineering Department, Hydraulics Division, Maslak, Istanbul, Turkey

Olga Kresse
Schlumberger, Sugar Land, USA

Xiaowei Weng
Schlumberger, Sugar Land, USA

Oagile Dikinya
Department of Environmental Science, University of Botswana, Botswana

Laurence Jouniaux
Institut de Physique du Globe de Strasbourg, Université de Strasbourg, Strasbourg France

Ali Rasoulzadeh
Water Engineering Dept., College of Agriculture University of Mohaghegh Ardabili, Ardabil Iran

Preface

Hydraulics is a topic in applied science and engineering dealing with the mechanical properties of liquids or fluids. At a very basic level, hydraulics is the liquid version of pneumatics. Fluid mechanics provides the theoretical foundation for hydraulics, which focuses on the engineering uses of fluid properties. In fluid power, hydraulics are used for the generation, control, and transmission of power by the use of pressurized liquids. Hydraulic topics range through some part of science and most of engineering modules, and cover concepts such as pipe flow, dam design, fluidics and fluid control circuitry, pumps, turbines, hydropower, computational fluid dynamics, flow measurement, river channel behaviour and erosion. The text *Understanding Hydraulics* introduces the fundamental concepts involved in hydraulics. First chapter focuses on unsaturated hydraulic conductivity of fractal-textured soils. In this chapter, the soil-water characteristic curve (SWCC) and relative hydraulic conductivity (RHC) function were derived and expressed by the effective degree of saturation based on the fractal model for the pore surface. The proposed soil–water characteristic curve and unsaturated hydraulic conductivity function were examined in detail. The impact of ice cover on local scour around bridge piers is presented in second chapter. Third chapter explores how changes in hydraulic conductivity may affect modelled rates of water flow through forested watersheds, with flows referring to infiltration, percolation, run-off, interflow, base flow, and stream discharge. Fourth chapter examines five limitations of hydraulic fractures and interpretation techniques, and describe the increases in well productivity that can be achieved when efforts are made to address and compensate for these deficiencies. Fifth chapter presents hydraulic and sleeve fracturing laboratory experiments on six rock types. Sixth chapter highlights on hydraulic fracturing in jointed rock. Seventh chapter presents the measured hydraulic conductivity results and the relationship among the hydraulic conductivity, RQD, DI, GCD, and LPI. The application of the proposed HC model was also addressed. The objective of eighth chapter is to derive hydraulic geometry relations using bed load formulae and compare them. Hydraulic fracturing in formations with permeable natural fractures has been presented in ninth chapter. Hydraulic conductivity of semi-quasi stable soils is presented in tenth chapter. Last chapter proposes a comprehensive review of the electrokinetic coupling in rocks and sediments and a comprehensive review of the different approaches to deduce hydraulic properties in various contexts.

Chapter 1

UNSATURATED HYDRAULIC CONDUCTIVITY OF FRACTAL-TEXTURED SOILS

Yongfu Xu

Department of Civil Engineering, Shanghai Jiao Tong University, Shanghai, China

INTRODUCTION

The increasing concern with groundwater pollution and contamination of soils has stimulated the development of numerous mathematical models of pollutant transport in soils. The most important approaches to model transient water and solute transport in the vadose zone are based on the Richards equation. To solve this equation, the knowledge of the soil hydraulic properties, namely, the soil-water characteristic curve (SWCC) and the unsaturated hydraulic conductivity is required. The laboratory measurements show that the value of unsaturated hydraulic conductivity varies considerably from soil to soil with different water content (Khaleel and Relyea, 1995). Indeed, it is found that the unsaturated hydraulic conductivity decreases by one to three orders of magnitude across a small pressure head range even near saturation (0~10cm pressure head), due to the effects of structural macropores (Jarvis and Messing, 1995). Of all hydraulic properties, the unsaturated hydraulic conductivity is most difficult to measure. Therefore the use of indirect methods has become more and more common to estimate the unsaturated hydraulic conductivity from more easily measured soil properties (van Genuchten et al., 1992).

Modern hydrological models require information on hydraulic conductivity and soil-water retention characteristics. All hydraulic properties, the soil-water characteristics, hydraulic conductivity and soil-water diffusivity (SWD) are closely related to the geometry of a porous media (Brooks and Corey, 1966; Burdine, 1953). Measurements of hydraulic properties are expensive, time-consuming and highly variable (Dirksen, 1991). Prediction of these properties is a viable alternative, especially when the predictive model contains a few parameters sensitive to structural conditions. Porous media (e.g. soils, rocks, etc.) are heterogeneous systems composed of numerous, different and interacting components (van Damme, 1995). The complex nature of these porous media complicates any prediction of their hydraulic properties.

A potentially powerful method results from the pore surface models, in which the soil–water characteristic curve of unsaturated soils is interpreted as statistical measure of its equivalent pore-size distribution (PSD) (van Genuchten, 1980; Corey, 1992). The frequency of different pore radii is related to matric suction by the soil-water characteristic curve (SWCC) and the relation between matric suction and pore radius is described by the Young-Laplace equation. The relative hydraulic conductivity (RHC) of unsaturated soils can be deduced from the soil-water characteristic curve (SWCC) through simplifying assumptions on pore topology and using the Hagen-Poiseuille law as an approximation for water flow. An additional empirical parameter is introduced, which includes all uncertainties. This empirical parameter is often referred to as "tortuosity", but its physical meaning is unclear (Vogel and Roth, 1998). The choice of the analytical model for the soil-water characteristic curve (SWCC) can significantly affect the predicted function of the unsaturated hydraulic conductivity.

Fractals describe hierarchical systems and are suitable to model the heterogeneous soil structure with tortuous pore space (Rieu and Sposito,1991; Xu and Sun, 2002). Toledo et al. (1990) modeled the soil-water characteristic curve (SWCC) and unsaturated hydraulic conductivity using fractal geometry and thin-film physics. Tyler and Wheatcraft (1990) gave the unsaturated hydraulic conductivity functions based on the fractal model for the soil-water characteristic curve (SWCC) and the relative conductive models developed by Mualem (1976) and Burdine (1953). Crawford (1994) studied the influence of heterogeneity of both the solid matrix and the pore space, as well as the shape of the pore boundary, on the saturated and unsaturated hydraulic conductivity. Fuentes et al. (1996) derived an expression for unsaturated hydraulic conductivity using the fractal dimension obtained from the soil-water characteristic curve (SWCC). Hunt and Gee (2002) applied critical path analysis from percolation theory to calculate the unsaturated hydraulic conductivity of soils with fractal pore space. Xu (2004),Xu and Dong (2004) Xu et al. (2004) derived the unsaturated hydraulic conductivity using the fractal model for the pore surface and gave a simple method to determine the fractal dimension. Models of the unsaturated hydraulic conductivity incorporate fractal dimension characterizing scaling of different properties including parameters representing connectivity and tortuosity. Thus, it is encouraged to derive the functions of the soil–water characteristic curve and unsaturated hydraulic conductivity from the fractal model for the pore surface. In this chapter, the soil-water characteristic curve (SWCC) and relative hydraulic conductivity (RHC) function were derived and expressed by the effective degree of saturation based on the fractal model for the pore surface. The proposed soil–water characteristic curve and unsaturated hydraulic conductivity function were examined in detail by the published experimental data. Comparisons between the prediction using both the fractal model and the van Genuchten-Mualem (G-M) model and the measurement of the relative hydraulic conductivity (RHC) were conducted. Using the fractal model for the soil-water characteristic curve (SWCC) and relatively hydraulic conductivity (RHC), one-dimensional rainfall infiltration and slope stability due to rainfall infiltration were analysized in chapter.

FRACTALS

Some of the most common and useful examples of fractals are: the Koch curve, the Sierpinski gasket and carpet, and the Menger sponge. All of these fractals enjoy the property of self-similarity. Roughly speaking, a subset of $\Re^n$ is said to be self-similar if it is a union of a number of smaller similar copies of itself. Some of the essential notions of the theory of fractals as it applies to self-similar sets should be briefly reviewed. A previous acquaintance with at least one of the sets mentioned above is useful but not necessary. Given an arbitrary subset B of $\Re^n$, a cover of B is a family U of sets $U_0 \subset \Re^n$

such that:

$$B \subset \bigcup_n U_\alpha$$

(1)

If the family U is countable (or finite) the cover is said to be countable (or finite). It is customary in that case to indicate the members of the family with a Latin subscript, namely, U_i, where i ranges over the natural numbers. The diameter of a subset of $\Re_n$ is the least upper bound (i.e., the supremum) of the distance between pairs of points in the subset. By convention, the diameter of the empty set is zero. A δ-cover of $B \subset \Re^n$ is a (countable) cover such that, for every i, $\text{diam}(U_i) \leq \delta$, where δ is a positive real number and where diam(.) is the diameter function on subsets of $\Re^n$. The members of a δ-cover are indicated by U_i^δ. The s-dimensional Hausdorff measure $H_s(B)$ of B is defined as:

$$H^s(B) = \lim_{s \to 0} \inf \sum_{i=1}^{n} [\text{diam}(U_i^s)]^s$$

(2)

where the inf extends over all δ-covers. It can be shown that this definition indeed provides an outer measure for all non-negative values of s and for any subset of $\Re^n$. The Hausdorff dimension of B is defined as:

$$\dim_H B = \inf\{s \geq 0 : H^s(B) = 0\}$$

$$(3)$$

The Hausdorff dimension of a subset of $\Re^n$ cannot exceed n. If $\dim_H B > 0$, then it can be shown that for all values of s strictly smaller (respectively, larger) than the dimension, the s-dimensional Hausdorff measure of B is infinite (respectively, zero). In this sense, the Hausdorff dimension represents a critical value of discontinuity of the s-dimensional Hausdorff measure, thus making Eq. (3) meaningful. Of particular interest is the value of the s-dimensional Hausdorff measure for s equal to the Hausdorff dimension of the set. This is usually called the Hausdorff measure of the set. The Hausdorff measure of a set may turn out to have any value, including zero or infinite.

An important feature of the s-dimensional Hausdorff measure is the scaling property, namely, the way it changes under similarity transformations of $\Re^n$. Recall that a transformation S: $\Re^n \to \Re^n$ is called a similarity if there exists a real scale factor $\lambda > 0$ such that, for all x, y $\in \Re^n$, the following equation is satisfied:

$$\left|S(y) - S(x)\right| = \lambda\left|y - x\right|$$

$$(4)$$

where $|.|$ denotes the Euclidean distance in $\Re^n$. It is not difficult to prove that for all subsets $B \subset \Re^n$ and for all values of s, under a similarity transformation S with scale factor λ the s-dimensional Hausdorff measure transforms according to the formula:

$$H^s(S(B)) = \lambda^s H^s(B)$$

$$(5)$$

Another important property of the s-dimensional Hausdorff measure is that it is preserved under Euclidean isometries (translations, rotations, reflections). In other words, congruent sets have the same s-dimensional Hausdorff measure. The behavior of the Hausdorff measure under general affine transformations, on the other hand, cannot be captured under the umbrella of a simple formula.

The properties just described of the s-dimensional Hausdorff measure can be used to obtain a straightforward evaluation of the Hausdorff dimension of self-similar fractals. Indeed, let B be the union of m copies of λ-scaled mutually congruent copies of itself (with $\lambda < 1$). If these copies are disjoint Borel sets, we have:

$$H^s(B) = m\lambda^s H^s(B)$$

$$(6)$$

by virtue of the scaling property (5). Assume now that the Hausdorff measure of B (namely, the value of $H_s(B)$ for s=dim H(B)) is finite and positive. In that case, it follows Eq. (6) that:

$$\dim_{H} B = -\frac{\log m}{\log \lambda} \tag{7}$$

The fractal dimension for a pore surface can be defined in the following way. Imagine the pores being enclosed by a set of spheres of radius r, the number of spheres N necessary to do this is clearly a function of radius of the spheres. The definition of the fractal dimension D is:

$$D = \lim_{r \to 0} \left[\frac{\ln(N(A,r))}{\ln(1/r)} \right] \tag{8}$$

Equation (8) is often used to determine the fractal dimension by experiments. A typical fractal set, the middle third Cantor set may be constructed from a unit interval by a sequence of deletion operations (Fig. 1). Let E_0 be the interval $[0,1]$, E_1 is the set obtained by deleting the middle third of E_0 so E_1 consists of the two intervals $\left[0, \frac{1}{3}\right], \left[\frac{2}{3}, 1\right]$. Deleting the middle third of E_1 gives E_2, thus E_2 comprises the four intervals $\left[0, \frac{1}{9}\right]\left[\frac{2}{9}, \frac{3}{9}\right]\left[\frac{6}{9}, \frac{7}{9}\right]\left[\frac{8}{9}, \frac{9}{9}\right]$. Proceeding in this like manner, E_i is obtained by deleting the middle third of each interval in E_{i-1} and Ei consists of 2^i intervals with length of 3^{-i}. Using Eq. (8), the fractal dimension of the middle-third Cantor set is $D = \lim_{i \to \infty} \left| \frac{\log(2^i)}{\log(1/(3^{-i}))} \right| = 0.63$. The Cantor set is often used to model the dust distribution.

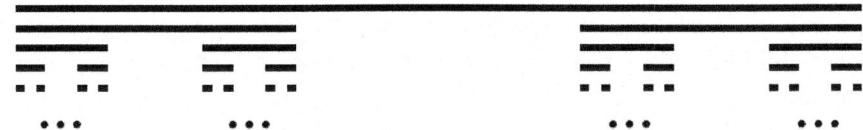

Figure 1. Middle-third Cantor set with fractal dimension D=0.63.

FRACTAL MODEL FOR THE PORE SURFACE

Many research results show that the soil pore surface is fractal (Avnir and Jaroniec 1989; Xu and Sun 2002). To investigate the impacts of fractal scaling upon hydraulic properties of porous media, a fractal representation of a porous media is developed. A cube of size 1 by 1 is formed by initially subdividing the cube into $1/a^3$ subcubes, each of size a by a. From the original cube, N subcubes are removed and represent a pore of size a by a. The remaining $1/a^3$-N subcubes are then each divided into $1/a^3$ subcubes and N subcubes are removed from each of the original subcubes. Such a recursion algorithm results in a cube everywhere filled with pores of all sizes, with a predominance of small pores. The number of size of a^3 needed to cover the pores equal to or larger than a^3 is given by N. The Menger sponge (Fig. 2) is often used to model porous materials, such as soils. For the Menger sponge, a=1/3, N=7, and the fractal dimension is given by

$$D = \lim_{i \to \infty} \left\{ \frac{\log(20^i)}{\log(1/(3^{-i}))} \right\} = 2.73.$$

Figure 2. The Menger sponge with fractal dimension D=2.73.

When we covered a soil pore surface using balls with the same radius, the soil pore surface is covered by N_1 balls with radius r_1, the same surface is covered by N_i balls with radius r_i, and it is covered by N_j balls with radius r_j, and usually $r_1 > r_i > r_j$, it is found that the surface area of the soil pore surface contents has the following relationship: $N_0 r_0^2 < N_i r_i^2 < N_j r_j^2$. If the soil pore surface is a self-similar surface, i.e., the pore surface is a fractal surface, the relationship between the number of covered balls and its radius can be written as (Mandelbrot 1982)

$$N(r) = Cr^{-D}$$

$$(9)$$

where C is a constant, r is the size corresponding to the pore radius, D is the fractal dimension of the pore surface. All materials have a surface fractal dimension in the whole range of physically meaningful values, i.e., from 2 to 3. The limit D=2 would correspond to a perfectly regular, smooth (Euclidean) surface, whereas D=3 would correspond to a self-similar surface so intricate that it would be space filling.

The fractal dimension of the pore surface can be determined using mercury intrusion porosimetry and adsorption isotherm. These two methods are introduced in detail as follows.

Mercury Intrusion Porosimetry

The d-measure is obtained from Eq. (9), and is expressed as follows:

$$M(r) = N(r)r^d = Cr^{d-D}$$

(10)

For the fractal pore surface, the relationship between the surface area with the radius less than r and the pore radius r can be obtained from Eq. (10). The surface area can written as

$$A_p(\leq r) = Cr^{2-D}a)$$
$$V_p(\leq r) = Cr^{3-D}b)$$

(11)

where in 11 a) $A_p(\leq r)$ is the surface area of the soil pores with the radius smaller than r. Similarly, the total volume of the pores with the radius less than r is given by 11b), where in $V_p(\leq r)$ is the cumulative volume of the soil pores with the radius smaller than r.

Neimark (1992) gave a method to measure the surface fractal dimension of the soil pores using mercury intrusion porosimetry. The soil pore surface can be approximated by the equilibrium interface between mercury and soil particles in the close vicinity of the pore surface. According to the Young–Laplace equation, the mean radius of the equilibrium interface is expressed as follows:

$$r(p) = \frac{2\sigma \cos \alpha}{p}$$

(12)

where r(p) is the mean radius of the equilibrium interface under pressure p, σ is surface tension. From the thermodynamic viewpoint, the equilibrium interface area can be calculated from the balance between the work of formation of the equilibrium interface and the work of mercury intrusion (Neimark 1992), i.e.,

$$A_p(p) = \int_0^{V(p)} \frac{V(p)}{T} dp$$

(13)

where $A_p(p)$ is the equilibrium interface area under pressure p. The surface fractal dimension can be determined theoretically from Eq. (11a). If the slope of the straight line is λ in log r(p) versus log $A_p(p)$, the surface fractal dimension of soil pores is given by

$$D = 2 - \lambda$$

(14)

The pore-size distribution (PSD) is usually obtained using mercury intrusion porosimetry. According to Eq.(11b), the fractal dimension of the pore surface can be obtained from the slope of the regressed linear relation in the plane of logr vs. logV$_p$. If the slope of the fitting line in the logr-logV$_p$ plane is κ, the fractal dimension of the pore surface is written as

$$D = 3 - \kappa \tag{15}$$

The value of fractal dimension spans a large range from 1.0 to 3.0 (Gimenez et al., 1997). Larger fractal dimension is associated with clayey soils (van Damme, 1995).

The accumulative volume of the pores can be measured by many methods. Mercury intrusion porosimetry provides a useful and potentially valuable measurement of the pore volume for the porous medium. The ranges of equivalent pore diameter explored cover almost five orders of magnitude, from several hundred microns down to approximately 16Å (Watabe et al., 2000).

Figure 3 shows the soil pore-size distribution (PSD) obtained from mercury intrusion porosimetry (Watabe et al., 2000). Symbols S-02, S-03 and S-04 are the serial numbers of soil samples, and V_v is the total pore volume in Fig. 3. It is seen from Fig. 3 that the surface of the soil pores can be described by fractal model, and the relationship between $V_p(\leq r)$ and r can be expressed by a linear function in log-log plot. The slopes of the regressed linear relation in the plane of logr vs. $\log(V_p(\leq r)/V_v)$ are 0.34, 0.37 and 0.49, and therefore the fractal dimensions are 2.66, 2.63 and 2.51 for specimens of S-02, S-03 and S-04, respectively. The maximum radius is the radius at which the pore volume reaches the maximum value, and is defined as the intersection between the fitting line and the line of $V_p(\leq r)/V_v = 100\%$. The maximum radii of soil pores are 0.0125, 0.06 and 0.006 mm for specimens of S-02, S-03 and S-04, respectively. The relationship between the maximum radius R and the air-entry value ψ_e can be expressed by the Young-Laplace equation, i.e. $\psi_e = 2\sigma\cos\alpha/R$. The parameters $\alpha=0$ and $\sigma=0.075$kPa mm (Watabe et al., 2000), and the air-entry values are 12kPa, 2.5kPa and 25kPa corresponding to the maximum pore radius R of 0.0125, 0.06 and 0.006 mm for specimens of S-02, S-03 and S-04, respectively. The parameters obtained from the fractal model of the pore surface are listed in Table 1. InTable 1, parameters R and κ are obtained from the fractal model of the pore surface in Fig. 3.

Table 1. Parameters obtained from the fractal model of PSD

Soil type	κ	D	δ	*R (mm)*	ψ_e*(kPa)*	Data source
S-02	0.34	2.66	-0.34	0.0125	12	Watabe et al., 2000
S-03	0.37	2.63	-0.37	0.06	2.5	
S-04	0.49	2.51	-0.49	0.006	25	
A horizon	0.22	2.78	-0.22	0.2	0.75	Vogel and Roth, 1998
B horizon	0.13	2.87	-0.13	0.25	0.6	
Toyoura sand	1.33	1.67	-1.33	0.09	1.67	Uno et al., 1998

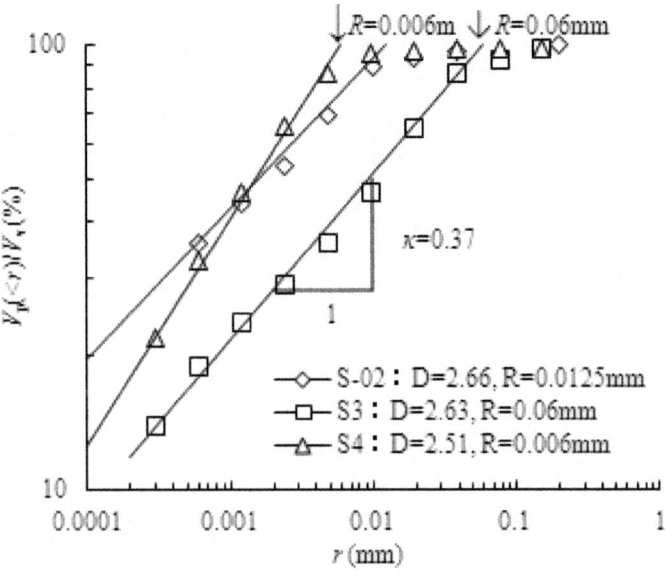

Figure 3. Fractal dimension and the maximum pore radius obtained from PSD for glacial tills (Data from Watabe et al., 2000)

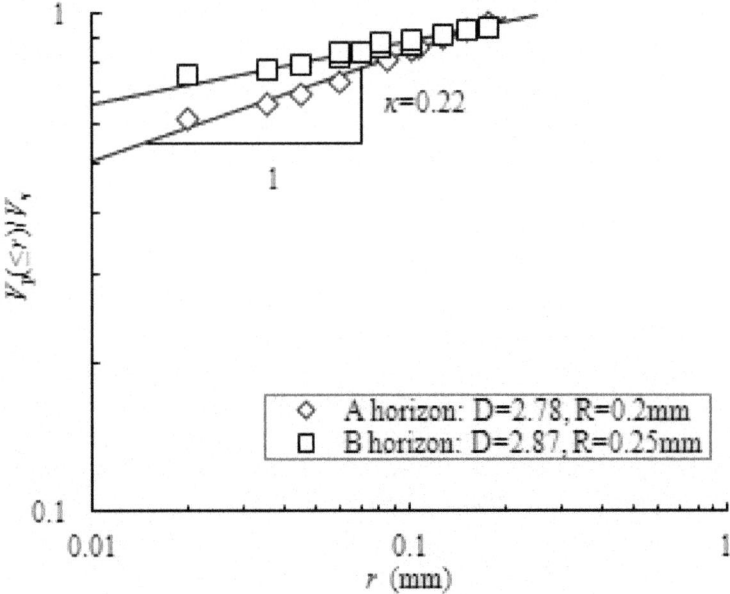

Figure 4. fig. 4 Fractal dimension and the maximum radius of the PSD of two silty soils (Data from Vogel and Roth, 1998).

Experimental data of the pore-size distribution (PSD) for agricultural silty soils from two horizons, A and B were also taken from Vogel and Roth (1998). The pore-size distribution (PSD) was determined using the three-dimensional reconstructions of the pore space. The pore space was eroded by a spherical structuring element with a given diameter 2r and subsequently dilated by the same structuring element (Vogel and Roth, 1998). In the resulting image all pores smaller than 2r were removed. The cumulative volume $V_p(>r)/V_T$ was obtained using the application of successively larger structuring elements byVogel and Roth (1998). The following relationship is obtained between $V_p(>r)/V_T$ and $V_p(\leq r)/V_T$.

$$\frac{V(>r)}{V_T} + \frac{V(\leq r)}{V_T} = \frac{V_v}{V_T} = \theta_s$$

(16)

where V_T is the total volume of soil, V_v is the total volume of void, θ_s is the saturated volumetric water content. The saturated volumetric water contents are 0.502 and 0.448 for the soils at A horizon and B horizon, respectively. Hence the cumulative volume $V(\leq r)/V_T$ can be obtained from the cumulative volume $V(>r)/V_T$. The cumulative volume of $V(\leq r)/V_T$ can be translated into $V(\leq r)/V_v$ according to Eq.(16). The cumulative volume of $V(\leq r)/V_v$ is shown in Fig. 4, where the dots denote the results transformed from the experimental data given by Vogel and Roth (1998) and the solid lines denote the results of the fractal model. The relationship between the pore volume and the pore radius satisfies an approximately linear expression in log-log plot. The slopes (κ) of the regression line are 0.24 and 0.13 for the soils at A and B horizons, respectively. According to Eq. (15), the fractal dimensions of the pore surface are 2.76 and 2.87 for A and B horizons, respectively. The maximum radius is defined as the value at which the cumulative volume of pores reaches the maximum, and $V(\leq r)/V_v = 100\%$. The maximum radii of the soil pores at the A and B horizons are 0.2cm and 0.25cm, respectively. The air-entry values can be calculated using the Young-Laplace equation, and they are 0.75kPa and 0.6kPa for the A and B horizons, respectively. The fractal dimension of the soil pore and the air-entry value of the soils at the A and B horizons are listed in Table 1. It can be seen from Figs. 3-4 that the pore surface can be modeled by fractal theory, and the pore-size distribution (PSD) satisfies with Eq. (11b). The soil structure can be described by the fractal dimension. The larger the fractal dimension, the more tortuous is the pore space. The fractal dimension is varied with the soil structure, and is different for different soils.

Uno et al. (1998) studied the relationship between the pore-size distribution (PSD) and unsaturated hydraulic properties of Toyoura sand. The pore-size distribution (PSD) of Toyoura sand was measured by the air intrusion method, and is shown in Fig. 5. The relationship between the pore volume and pore radius is approximated by a linear function in log-log plot, and satisfies with Eq. (11b). Hence the pore-size distribution (PSD) of Toyoura sand can be described by fractal model, and its fractal dimension is 1.67 obtained from Fig. 5 not increase with radius, i.e., $V_p/V_v = 1$. The maximum pore radius of Toyoura sand is nearly 0.09 mm in Fig. 5. The surface tension (T) of water is 0.075kPa mm. The air-entry value can be calculated from Young-Laplace equation, and is 1.67kPa on the assumption that $\alpha=0$. The fractal dimension of the soil pore and the air-entry value of Toyoura sand are listed in Table 1.

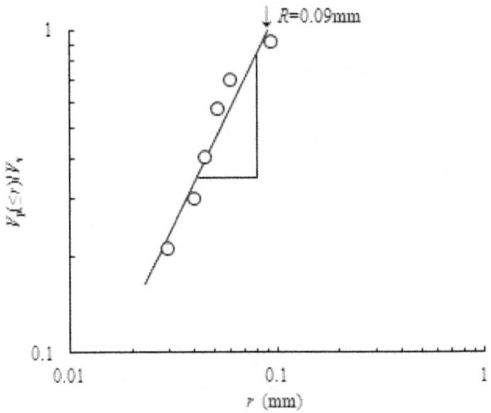

Figure 5. The fractal dimension and the maximum radius evaluated from the PSD of Toyoura sand. (Data fromUno et al., 1998.)

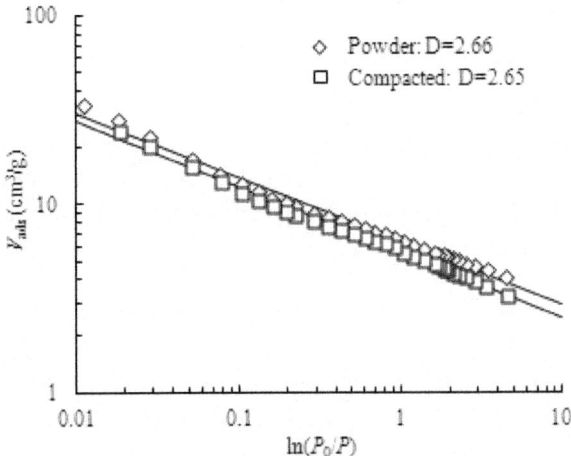

Figure 6. The surface fractal dimension of Tsukinuno bentonite from nitrogen adsorption.

Adsorption Isotherm

The fractal dimension of the pore surface serves to characterize the pore surface roughness or irregularities. The magnitude of the fractal dimensions is relevant to many important physicochemical processes namely adsorption, surface diffusion and catalysis. Avnir and Jaroniec (1989) proposed a convenient method to determine the fractal dimension from the nitrogen adsorption. Similarly with the adsorption isotherm equation, the correlation of the water volume absorbed by clay to the vapour pressure is given by

$$\frac{V_w}{V_m} = k\left[\ln\left(\frac{P_0}{P}\right)\right]^{D-3}$$

(17)

where V_w is the water volume absorbed by clay, V_m is the volume of montmorillonite, P is the partial water vapour pressure in equilibrium with clay at some water content w and some temperature T and P_0 is the equilibrium water vapour pressure of pure water at temperature T. The surface fractal dimension of Tsukinuno bentonite obtained from nitrogen adsorption is shown in Fig. 6. The surface fractal dimensions of bentonite powder and compacted bentonite are 2.66 and 2.65, respectively.

The adsorption isotherm also allows the calculation of swelling pressure of clay as a function of the water content (Kahr et al., 1990). Satisfactory agreement is found between the calculated and the experimental data for water content bigger than 10%.

$$p_s = -\frac{\overline{R}T}{M_w \nu_w}\ln\left(\frac{P_0}{P}\right)$$

(18)

where p_s is swelling pressure, the maximum axial pressure which is needed to maintain the original sample height, $\overline{R}$ is the molar gas constant, T is the Kelvin temperature, M_w is molecular mass of water and v_w is partial specific volume of water. Combining Eq. (17) with (18), the relationship between normalized water volume by the clay volume and swelling pressure is written as

$$\frac{V_w}{V_m} \propto p_s{}^{D-3}$$

(19)

The surface fractal dimension of clay can be obtained from Eq. (19). The experimental data of the swelling pressure and swelling deformation tests are compiled in Fig.7. It is seen that the relationship between the normalized water content and swelling pressure or vertical overburden pressure can be described by the same linear function which is expressed in Eq. (19). Therefore, Eq. (19) represents a general relationship between the water volume absorbed by clay per unit volume and vertical pressure supplied to specimen in swelling process. The surface fractal dimension of Tsukinuno bentonite is 2.63, which nearly equals to that obtained from Fig. 6. In Fig. 6, C_b is the bentonite content.

Swelling deformation tests of Wyoming bentonite were also conducted by Millins et al. (1996) and Studds et al. (1998). The relationship between the clay void ratio (e_c) and vertical pressure is shown inFig. 8. Relationship between the clay void ratio (e_c) and normalized water volume is given by

$$e_c = \frac{1}{C_m} \frac{V_w}{V_m}$$

(20)

where C_m is the montmorillonite content, and $C_m = 75\%$. It is seen that the slope of $\log e_c$-$\log p$ equals to that of $\log(V_w/V_m)$-$\log p$ from Eq. (20) for C_m being constant. Thus, the surface fractal dimension of Wyoming bentonite can be obtained from the linear relationship of $\log e_c$-$\log p$. The surface fractal dimensions of Wyoming bentonite are 2.64, which obtained from the swelling deformation tests conducted by different authors.

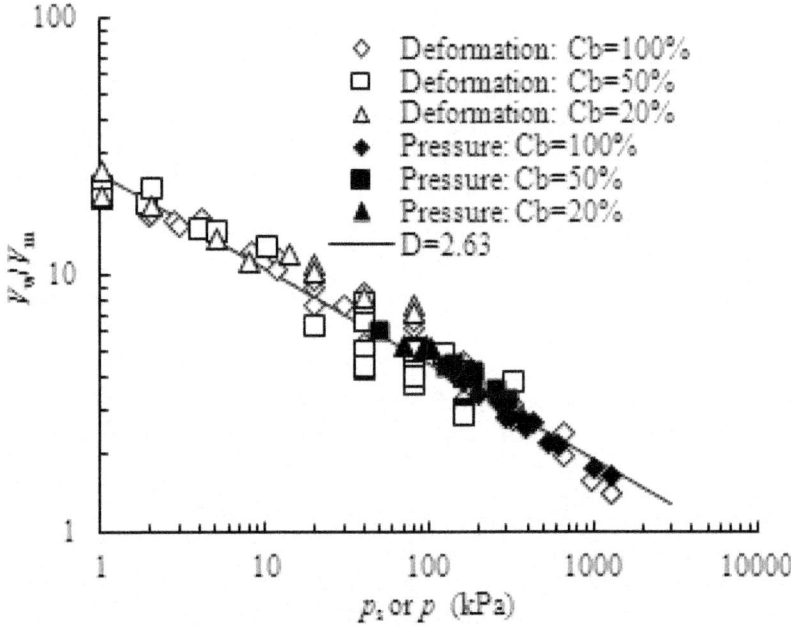

Figure 7. Relationship between absobed water volume and swelling pressure or vertical overburden pressure of Tsukinuno bentonite.

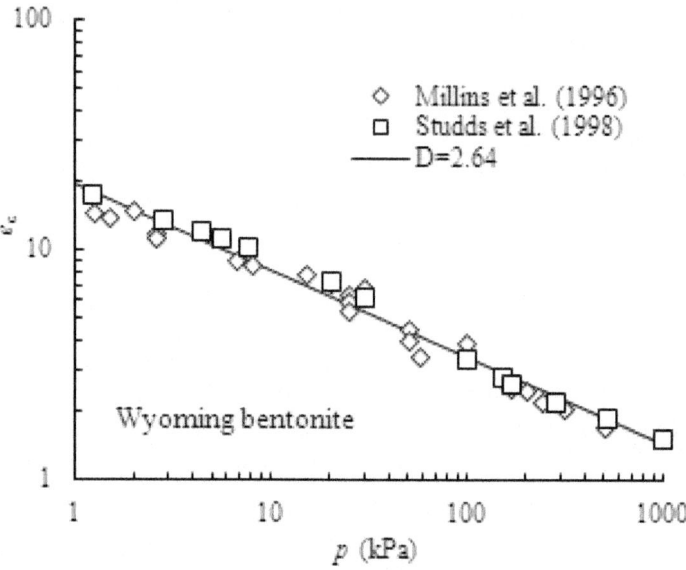

Figure 8. Relationship between clay void ratio and vertical overburden pressure of Wyoming bentonite.

Correlation between The Fractal Dimension Of PSD And That Of The GSD

The pore-size distribution (PSD) is not often and easily measured experimentally. The grain-size distribution (GSD) is easily measured and the fractal dimension of the pore surface can be conveniently obtained from the grain-size distribution (GSD). The relationship between the pore-size distribution (PSD) and the grain-size distribution (GSD) can be constructed for the soils with a homogeneous fabric. Consider a pore limited by a number of grains N_g, the pore radius r is related to the radius r_g of the smallest grain as follows (Watabe et al., 2000)

$$r = f_p r_g \tag{21}$$

where fp is a constant less than 1.0. If the particles are randomly arranged, the probability of having a grain of radius less than r_g is dP_g, the probability of having the N_g-1 grains with a radius greater than r_g is $(1-P_g)^{N_g-1}$. The probability P of having a pore with a radius less than r is given by (Watabe et al., 2000)

$$dP = N_g(1 - P_g)^{N_g-1} dP_g \tag{22}$$

It is assumed that the grain-size distribution (GSD) satisfies the fractal model, and is given by

$$P_g = (r_g/R_g)^{3-D_g} \tag{23}$$

where D_g is the fractal dimension of the grain-size distribution (GSD), R_g is the maximum radius of soil grains. The probability P is given by (Watabe et al., 2000)

$$P = 1 - \left[1 - \left(\frac{r}{f_p R_g}\right)^{3-D_g}\right]^{N_g} \tag{24}$$

If $r/(f_g R_g) \ll 1$, Eq.(24) can be written as

$$P = N_g \left(\frac{r}{f_p R_g}\right)^{3-D_g} \tag{25}$$

It is seen that the pore surface has the same fractal dimension as that obtained form the grain-size distribution (GSD) from Eqs. (23) and (24). Comparisons between the pore-size distribution (PSD) and the grain-size distribution (GSD) are shown in Fig. 9, where $P=V(\leq r)/V_v$ and $P_g=M(\leq r)/M_r$, V and M are the pore volume and grain mass, respectively. It is seen from Fig. 9 that both the pore-size distribution (PSD) and the grain-size distribution (GSD) can be expressed by fractal model, and they have the nearly same fractal dimension.

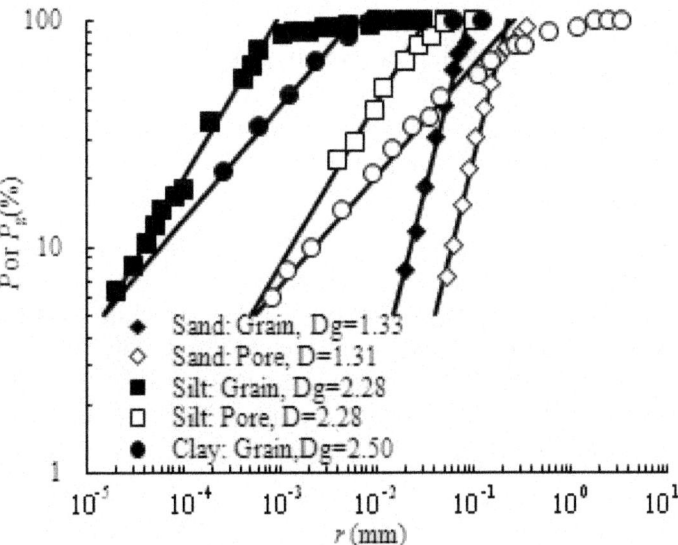

Figure 9. Correlation of the pore-size distribution (PSD) to the grain-size distribution (GSD).

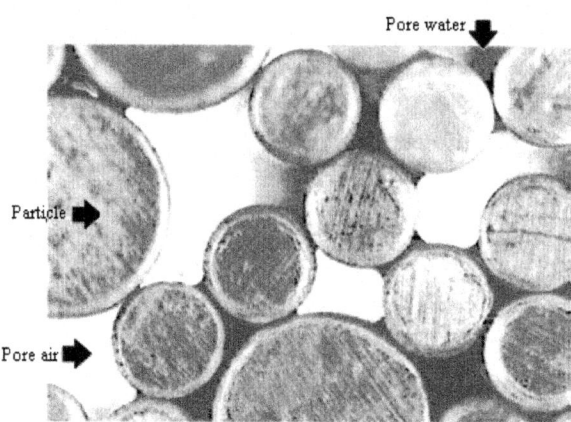

Figure 10. Model test for simulating the water distribution.

SOIL-WATER CHARACTERISTIC CURVES (SWCC)

The soil-water characteristic curve defines the relationship between water content (gravimetric water content w, volumetric water content θ) or degree of saturation (S) and matric suction (ψ). The soil water characteristic curve (SWCC) has been used extensively in the study of unsaturated soils and relates the water content and soil matric suction. The relationship between the matric suction and the radius of the incurvated surface between the pore-air and pore-water is expressed by the Young-Laplace equation. The hydraulic and mechanical properties of unsaturated soils are correlated to the microstructures of soils through the Young-Laplace equation.

For unsaturated soils, the pore water is usually localized in small micropores, and the micropores are usually fully filled with water. The water content in macropores is very little, sometimes, there is no water in macropores in highly unsaturated soils (Fig. 10). The water distribution (black domain) of model test shown in Fig. 10 proved the above assumptions. In the model test, the soil grains of different sizes were simulated by the aluminum rods with different radius (Matsuoka, 1999). The water mainly distributed in the small pore and no water exists in the large pore for unsaturated soils in the model test.

Three air-water distribution states were identified as saturated funicular state, complete pendular state, and partial pendular state, respectively. All the voids are filled with liquid in the saturated state. Air bubbles are present in the voids in the funicular state, and the liquid phase is continuous. There is no continuity in the water phase in the pendular state. Pore water within unsaturated soils can be divided into three forms (Wheeler and Karube, 1996)

(Fig. 11): bulk water within those void spaces that are completely flooded, meniscus water surrounding all inter-particle contact points that are not covered by bulk water and absorbed water (which is tightly bounded to the soil particles and acts as parts of the soil skeleton). The relationship between the soil-water characteristic curve (SWCC) and the pore-water forms is shown in Fig. 11. When the volumetric water content is less than its residual value, the pore-water is tightly absorbed by soil particle and cannot move freely. This absorbed water is seen as a part of soil particles. Thus, the contribution of the filled pores with radius r→dr to the water content is given by

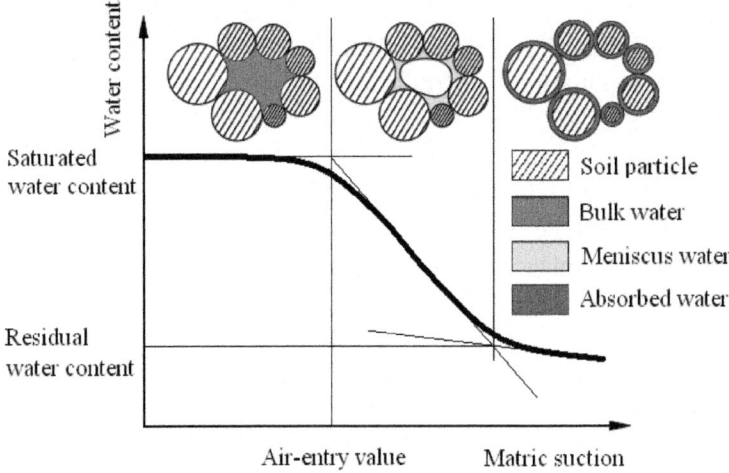

Figure 11. Relationship between the SWCCs and the pore-water forms of unsaturated soil (Matsuoka, 1999).

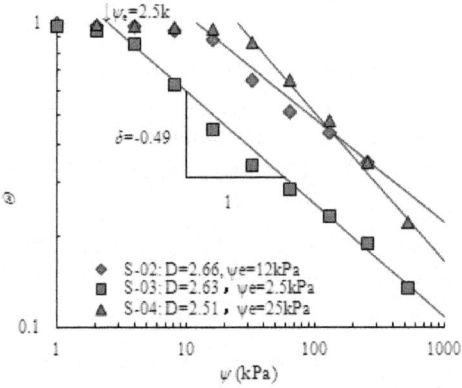

Figure 12. Comparison between the prediction and experiments of the SWCCs (Data from Watabe et al., 2000).

$$dA = \frac{N4\pi r^2 dr}{V_T}$$

(26)

where A is the relative volumetric water content, and $A = \theta - \theta_r$, θ and θ_r are the actual and residual volumetric water content, respectively. The residual volumetric water content is the volumetric water content at which the effectiveness of matric suction to cause further removal of water requires vapour migration. Substituting Eq. (9) into Eq. (26), A is given by

$$A = Br^{3-D}$$

(27)

where $B = 4\pi C/[V_T(3-D)]$. Similarly with Eq. (27), the relatively volumetric water content at saturation is written as follows

$$A_s = BR^{3-D}$$

(28)

where R is the maximum radius of soil pores.

The relationship between matric suction ψ and the pore radius is obtained from the Young-Laplace equation. Thus, the soil-water characteristic curve (SWCC) is obtained as follows

$$\Theta = \left(\frac{\psi}{\psi_e}\right)^{\delta}$$

(29)

where $\delta = D-3$, Θ is the normollized volumetric water content, and is written as follows:

$$\Theta = \frac{\theta - \theta_r}{\theta_s - \theta_r} = \frac{S - S_r}{100 - S_r}$$

(30)

where θ, θ_r and θ_s are the volumetric water content, residual volumetric water content and saturated volumetric water content, S and S_r are the degree of saturation and residual degree of saturation. The residual volumetric water content is not always made routinely, in which case it has to be estimated by extrapolating available soil-water characteristics data towards lower water content, such as shown in Fig. 11. van Genuchten (1980) defined the residual volumetric water content as the water content for which the gradient $d\theta/d\psi$ becomes zero at high matric suction. From a practical point of view it seemed sufficient to define θ_r as the water content at some large matric suction, e.g. at the permanent wilting point $\psi_e = 1500$kPa.

Equation (29) is the soil-water characteristic curve (SWCC) derived from the fractal model for the pore surface. The normalized volumetric water content is equal to unity at values of matric suction up to the air-entry value and equal to zero after residual saturation. The only two parameters ψ_e and D, which have obvious physical meaning, are used in Eq. (29) to express the soil-water characteristic curve (SWCC). The fractal dimension of the pore surface and the maximum pore radius can be evaluated from the pore-size distribution (PSD) measured using the mercury intrusion tests. The air-entry value can be calculated from the Young-Laplace equation using the maximum pore radius. Thus, the soil-water characteristic curve (SWCC) can be calculated from Eq. (29) using the fractal dimension and the air-entry value obtained from the mercury intrusion tests.

The soil-water characteristic curves (SWCC) were calculated from Eq. (29) using the fractal dimension of the pore surface and the air-entry value for specimens of S-02, S-03 and S-04, which listed in Table 1. The value of the normalized volumetric water content is calculated from Eq. (30). Here S_r=7.5% (Watabe et al., 2000). Experimental data show in good accord with the calculation of the soil-water characteristic curve (SWCC) in Fig. 12.

The experimental data of the soil-water characteristic curve (SWCC) measured in a multi-step outflow experiment were given by Vogel and Roth (1996) for silty soils. The saturated volumetric water contents are 0.502 and 0.448 for silty soils at the A and B horizons, respectively. The residual volumetric water content is equal 0 for both A and B horizons. Comparisons between the calculations ofEq. (29) and experimental data of the soil-water characteristic curves (SWCC) were shown in Fig. 13. The calculation of the soil-water characteristic curves (SWCCs) for silty soils was obtained from Eq. (29). The fractal dimensions of the soil pore and the air-entry values were listed in Table 1.

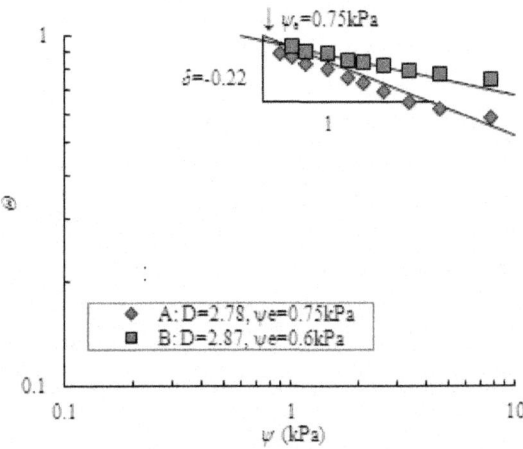

Figure 13. Comparison between prediction and experiments of SWCCs (Data from Vogel and Roth, 1996).

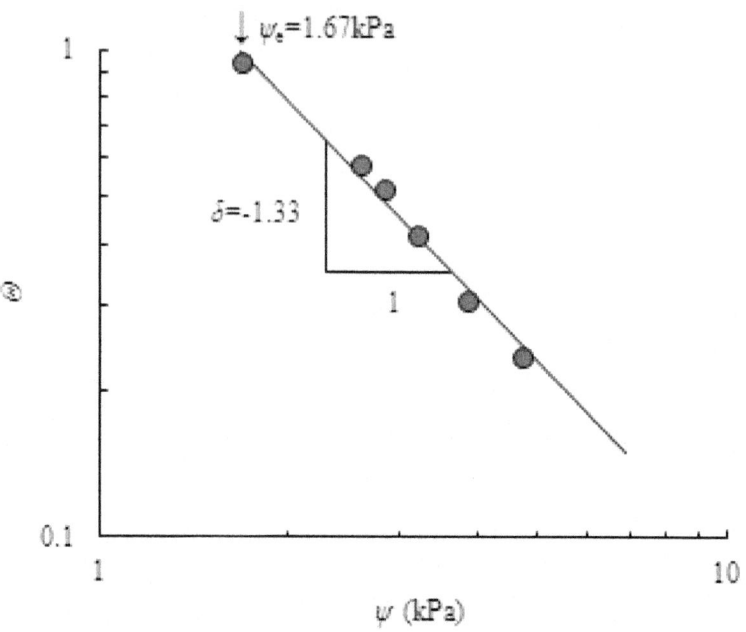

Figure 14. Comparison of the SWCCs for Toyoura sand (Data from Uno et al., 1998).

Uno et al. (1998) gave the experimental data of the soil water characteristic curves for Toyoura sand. Using the fractal dimension and the air-entry value obtained from the pore-size distribution (PSD), the soil-water characteristics can be simulated from Eq. (29) for Toyoura sand. The parameters obtained from the pore-size distribution (PSD) and used to predict unsaturated hydraulic conductivity are listed in Table 1. θ_s=0.425, θ_r=0. Comparison between the prediction of Eq. (29) and experimental data of the soil-water characteristics is shown in Fig. 14 for Toyoura sand.

Stingaciu et al. (2009) evaluated the feasibility of using nuclear magnetic resonance (NMR) relaxometry measurements to characterize pore size distribution and hydraulic properties in four porous samples with different texture and composition. The sandy samples FH31 and W3 were loaded in the penetrometer and packed at the same packing density as for the NMR measurements. The Mix8 and MZ samples were used as solid conglomerates. From the relaxation time distribution functions, the cumulative pore-size distribution functions were calculated with the average surface relaxivity. The normalized cumulative pore-size distributions functions were displayed in Fig. 15. The fractal dimension of pore surface and the maximum pore radius can be obtained from Fig. 15, and were listed in Table 2.

Table 2. Parameters for the prediction of SWCCs.

Sample	Method	Bulk density (g/cm³)	θ^s	θ^r	D	R (mm)	ψ_e(kPa)
FH31	Pressure plate	1.58	0.32	0.02	0.85	0.065	2.31
W3	MSO	1.42	0.33	0.058	1.51	0.0175	8.6
Mix8	Rosetta	1.45	0.41	0.061	1.52	0.00053	290
MZ	Pressure plate	1.60	0.44	0	2.29	0.00035	429

The soil-water characteristic curves (SWCC) were determined using the standard sand bed, pressure cell or multistep outflow method, and were plotted in Fig. 16. SWCC of FH31 and MZ were based on pressure plate measurements. SWCC of W3 was determined using multistep outflow (MSO) and that of Mix8 using ROSETTA software (Schaap et al., 2001). The matric suction as plotted in the abscissa (Fig. 16) was transformed into pore diameter using the Young-Laplace equation. In addition, the ordinate was the normalized volumetric water content. The predictions of SWCC were conducted usingEq. (29) with the parameters listed in Table 2. The parameters were derived from the pore-size distribution (PSD) shown in Fig. 15. Comparisons of the predictions and experimental data were shown in Fig. 16. The predictions of Eq. (29) were in good accord with the experimental data.

In general, the soil-water characteristic curves (SWCC) were usually interpreted as cumulative distribution functions in comparison to pore-size distribution (PSD) functions Using the soil-water characteristic curves (SWCC), pore-size distribution can be extracted for a given porous medium on the basis of an empirical law that related the pore suction to the effective pore radius. The soil-water characteristic curves (SWCC) and the pore-size distributions (PSD) are related by correlations as: (1) the absolute values of the slopes of SWCC and PSD are equal; (2) the air-entry value of SWCC is related to the maximum pore radius of PSD through the Young-Laplace equation.

The following conclusions can be obtained from Figs. 12-16: (1) The surface of the soil pore can be described by fractal model, and the surface fractal dimension of the soil pore and the air-entry value can be evaluated from the pore-size distribution (PSD). The soil-water characteristic curves (SWCCs) can be calculated using fractal model of the soil pore. (2) The fractal dimension and air-entry value obtained from the soil pore-size distribution (PSD) are equivalent to those obtained from the soil-water characteristic curve (SWCC). Hence the fractal dimension and air-entry value can be obtained from the fitting of the soil-water characteristic curve (SWCC) if the pore-size distribution (PSD) were not measured.

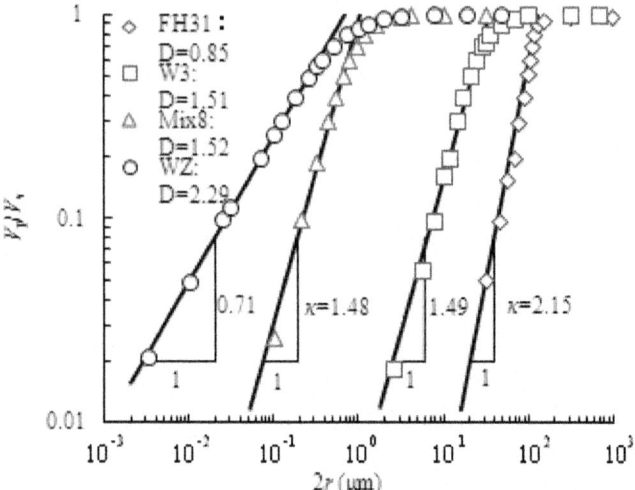

Figure 15. Cumulative pore-size distribution of for the three artificial substrates (FH31, W3, and Mix8) and the natural soil Merzenhausen (MZ), from which the fractal dimension of pore surface and the maximum pore radius were derived (Data from Stingaciu et al. 2009).

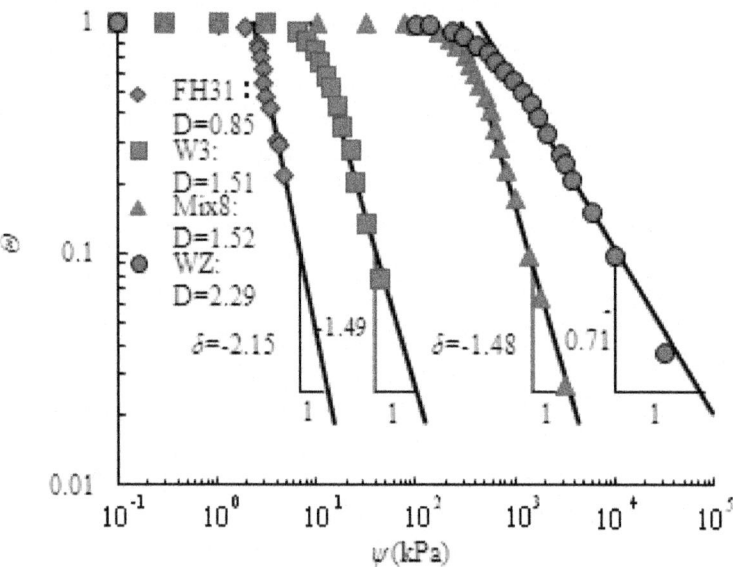

Figure 16. Comparison between the predictions of Eq. (29) and experimental data of SWCCs (Data fromStingaciu et al. 2009).

UNSATURATED HYDRAULIC CONDUCTIVITY AND DIFFUSIVITY

Brief Review

Brutsaert (2000) made a brief review of the capillary model, which was cited as follows.

It is very difficult to determine unsaturated hydraulic conductivity through experiments. Hydraulic conductivity for unsaturated soils is a function of soil-water content θ. For this reason, over the years many attempts have been made to represent the hydraulic conductivity by simple parametric equations, in terms of other properties of the soil which are easier to determine. One of these was of the following form (Brutsaert, 2000):

$$k_{\mathrm{r}} = \Theta^{\omega} \tag{31}$$

In the past, the power form of Eq. (31) has been derived on the basis of some widely different conceptual models of the pore geometry. There have been some marked differences in the obtained values of ω, depending on the adopted model of the pore geometry.

Uniform pore size models

A simple approach was that porous medium was characterized by some equivalent uniform pore size and the variability of the pore sizes was not considered. This approach invariably resulted in a constant value of ω. The capillary tube model of Averyanov (Polubarinova-Kochina, 1952) led to $\omega=3.5$, whereas the extension of the hydraulic radius model of Kozeny (Kozeny, 1927) to unsaturated soils byIrmay (1954) produced $\omega=3$.

Parallel models

In parallel models, the pore system was assumed to be equivalent with a bundle of uniform capillary tubes of many different sizes. The pore-size distribution was derived from the soil-water characteristics (SWCC), which can be related to the effective pore radius r by the Yong-Laplace equation. The true mean velocity in each pore was described by a Poiseuille-like equation for creeping flow. Purcell (1949) and Gates and Tempelaar-Lietz (1950) first used this approach. Several subsequent applications of the parallel model can be written in the common form (Brutsaert, 1967):

$$k_r = \frac{\beta \int_0^{\Theta} \psi(x)^{-2} dx}{\beta_0 \int_0^1 \psi(x)^{-2} dx}$$

(32)

where x is the dummy variable representing Θ. The variable $\beta = \beta(\Theta)$ is related to the tortuosity, and β_0 is its value at satiation, when $\Theta = 1$. The tortuosity concept was originally introduced by Carman (1956) as an improvement on the non-uniform hydraulic radius model of Kozeny (1927) and it can be expressed as $T = (L_e/L)^2$, in which L_e is the actual or microscopic path length of the fluid particles in the pores andL is their apparent or macroscopic path length along the Darcy stream lines. The parameter β was introduced in the parallel model by Wyllie and Spangler (1952), because without it, Eq. (32) yielded values considerably larger than their experimental data. Burdine (1953) proposed on the basis of his experimental data that $\beta/\beta_0 = \Theta^2$.

A further development was performed by adopting the following soil-water characteristic curve (SWCC) (Brooks and Corey, 1966):

$$\Theta = 1 \qquad \text{for} \psi \leq \psi_e$$
$$\Theta = \left(\frac{\psi}{\psi_e}\right)^b \qquad \text{for} \psi \leq \psi_e$$

(33)

Brooks and Corey (1966) integrated (32) with Burdine's assumption for β (Burdine, 1953) to obtain Eq. (31), with an exponent

$$\omega = 3 - \frac{2}{b}$$

(34)

It should be noted that, without Burdine's assumption for the tortuosity, the parallel model Eq. (32)applied with Eq. (33) would have yielded

$$\omega = 1 - \frac{2}{b}$$

(35)

Series-parallel models

The theoretical construction of this model also started with a bundle of parallel pores, each with a different but uniform size. However, these pores were then cut normally to the direction of flow with two resulting faces, and finally after some random rearrangement of the tubes the faces are joined again. This way account was taken of the random variations of the pore sizes, not only in the plane normal to the direction of flow, but also along the direction of flow. The discharge rate in each single pore, which consisted of two sections in cases, was assumed to be governed by the section with the smaller diameter. The pore-size distribution was derived form of the soil-water characteristics by means of the Yong-Laplace equation. The true velocity in each pore was obtained by means of a Poiseuille-like equation.

This model was originally proposed by Childs and Collis-George (1950) in a finite-difference scheme to calculate hydraulic conductivity from experimental data of the soil-water characteristics (SWCC). It was subsequently reformulated in integral form by Brutsaert (1968), to allow the derivation of more concise analytical expressions for hydraulic conductivity. The integral form can be written as

$$k_w = \left[\left(\frac{2T}{\gamma} \right)^2 \frac{\theta_0^2 (1-S_r)^2}{G} \right] \times \left[\int_0^\Theta \int_0^x \psi(x)^{-2} dy dx + \int_0^\Theta \psi(x)^{-2} \int_0^\Theta dy dx \right]$$

(36)

where G is a geometrical constant. In the case of Poiseuille's equation, G=8. It was trivial to show by integration by parts that the first double integral on the right was identical to the second. Hydraulic conductivity can be expressed concisely as (Brutsaert, 2000):

$$k_r = \frac{\int_0^\Theta (\Theta - x)\psi(x)^{-2} dx}{\int_0^1 (1 - x)\psi(x)^{-2} dx}$$

(37)

The capillary model similar with Eq. (37) in form was originally proposed for by Fatt and Dykstra (1951).

Fractal Model for Hydraulic Conductivity

Soil pore systems are viewed as a collection of interconnected voids. The individual voids are considered to have two equilibrium states: either filled by water or empty. Flow within the soil pores is assumed to be described by (Burdine, 1953; Mualem, 1976)

$$v = -\frac{r^2 g}{c\eta} \frac{dh}{dx}$$

(38)

where v is the average velocity within the pore; r is the radius of the pore; g is the gravity acceleration; η is the kinematic viscosity; c is a constant depending on the pore geometry; h is the hydraulic head, x is the position axis. Let us consider a porous slab of thickness Δx isolated from a homogenous soil column by two parallel cross-sections normal to the x axis. The areal porosity at each face of the slab is the same and it is equal to the volumetric porosity. The major simplifying assumption is that the actual configuration of slab may be replaced by a set of capillary tubes with different radii, parallel to the x axis. The flux dq flowing through the connections between pores of radius $r_1 \rightarrow r_1 + dr_1$ on one side of the slab (at x) and pores of radius $r_2 \rightarrow r_2 + dr_2$ on the other side of the slab (at x+dx) can be expressed by (Mualem, 1976)

$$dq = \beta r_e^2(r_1, r_2, \rho) A_e(r_1, r_2, \rho) \frac{dh}{dx} dr_1 dr_2$$

(39)

where β is a constant which incorporates the fluid with the matrix properties; $r_e(r_1, r_2, q)$ is the effective radius; $A_e(r_1, r_2, q)$ is the effective area; ρ is the maximum radius of the water filled pores at volumetric water content θ. According to the Darcy equation, the hydraulic conductivity k_w is given by

$$k_w = \frac{q}{dh/dx} = \beta \int_0^\rho \int_0^\rho r_e^2(r_1, r_2, \rho) A_e(r_1, r_2, \rho) dr_1 dr_2$$

(40)

The relatively hydraulic conductivity (kr) is the ratio of the hydraulic conductivity at any volumetric water content (kw) to the hydraulic conductivity at saturation (ks). The relative hydraulic conductivity (RHC) is related to the radius of the soil pores, and is written as (Mualem, 1976)

$$k_r = \frac{k_w}{k_s} = \frac{\int_0^r \int_0^r r_e^2(r_1, r_2, r) A_e(r_1, r_2, r) dr_1 dr_2}{\int_0^R \int_0^R r_e^2(r_1, r_2, r) A_e(r_1, r_2, r) dr_1 dr_2}$$

(41)

where r is the maximum radius of the water filled pores at the volumetric water content θ. It is assumed that the pore distribution at the two cross-section is completely random. The probability of pores of radius $r_1 \rightarrow r_1 + dr_1$ at x to encounter pores of radius $r_2 \rightarrow r_2 + dr_2$ at x+dx is given by

$$A_e(r_1, r_2) dr_1 dr_2 = f(r_1) f(r_2) dr_1 dr_2$$

(42)

The effective radius is assumed to be equal to the radius r reduced by a factor, which is equal to the ratio between the effective area to flow and the actual pore area. The effective radius is given by (Mualem, 1976)

$$r_e = r \frac{\int_0^r \int_0^r A_e(r_1, r_2, r) dr_1 dr_2}{\int_0^R f(r_1) dr_1} = r \Lambda^{1/2}$$

(43)

Substituting Young-Laplace equation and Eqs.(42) and (43) into Eq.(41), the relative hydraulic conductivity (RHC) is obtained as

$$k_r = \left(\frac{\psi}{\psi_e} \right)^\xi$$

(44)

where $\xi = 3D-11$. Substituting Eq. (29) into Eq. (44), the relative hydraulic conductivity (RHC) unction expressed by effective degree of saturation is written as follows

$$k_r = \Theta^{\xi/\delta} \tag{45}$$

Equations (44) and (45) are the relative hydraulic conductivity (RHC) function derived from the fractal model of the pore surface. The parameters in the proposed hydraulic conductivity function have an obvious physical implication, and can be determined from the pore-size distribution (PSD). Eqs. (44)and (45) offer a powerful, physical-based method to calculate the relative hydraulic conductivity (RHC). Eq. (44) is nearly similar with the Brooks and Corey equation (Brooks and Corey, 1966) in form, and the parameters in Eq. (44) have obvious physical implication. Eqs. (44) and (45) offer a powerful, physical-based method to calculate the relative hydraulic conductivity (RHC).

The expression of the soil-water diffusivity (SWD) can be derived from the unsaturated hydraulic conductivity and soil-water characteristics, and written as follows:

$$d = k_w \left| \frac{d\psi}{d\theta} \right| \tag{46}$$

where d is the soil-water diffusivity (SWD), k_w is the saturated hydraulic conductivity. SubstitutingEqs. (44) and (45) into Eq. (46), the soil-water diffusivity (SWD) is written as

$$d = k_s \psi_e \left(\frac{\psi}{\psi_e} \right)^{\varsigma} \tag{47}$$

$$d = k_s \psi_e \Theta^{\varsigma/\delta} \tag{48}$$

where $\varsigma = 2D - 7$.

As given in Eqs. (29), (44), (45), (47) and (48), methods to evaluate the soil-water characteristics (SWCCs), unsaturated hydraulic conductivity (RHC) and soil-water diffusivity (SWD) are proposed using the fractal model for the pore surface. The fractal dimension and the maximum radius of soil pores can be obtained from the pore-size distribution (PSD). The air-entry value can be calculated from the Young-Laplace equation using the maximum radius of soil pores. The soil-water characteristic curves (SWCC), unsaturated hydraulic conductivity (RHC) and soil-water diffusivity (SWD) can be predicted using the fractal dimension and air-entry value, which can be determined from the pore-size distribution (PSD). According to Eq. (29), the fractal dimension of the pore surface and the air-entry value can also be evaluated from the fitted soil-water characteristic curve (SWCC). Thus, the unsaturated hydraulic conductivity (RHC) and soil-water diffusivity (SWD) can also be estimated fromEqs. (44), (45), (47) and (48) using the fractal dimension and air-entry value evaluated from the soil-water characteristic curve (SWCC).

Predictions of SWCC and RHC from PSD

Vogel and Roth (1998) inferred effective hydraulic properties of unsaturated soil from the structure of the pore surface. The water retention characteristic and the hydraulic conductivity were simulated by network models with $32^3=32768$ nodes. The hydraulic conductivity, k_w, is determined at each step of desaturation by imposing a pressure gradient Δp across the ends of the network. Water flow q_{ij} in a bond between the nodes i and j through a horizontal surface A is described by Poiseuille's law. The hydraulic conductivity was calculated as $=qL/(A\Delta p)$, where L is the vertical length of the network. Comparison between the predictions of Eq. (44) and experimental data given by Vogel and Roth (1998) was shown in Fig. 17. The parameters used to do prediction were listed in Table 1, which were obtained from the fractal model for the pore-size distribution (PSD). The predictions were in good accord with the experimental data.

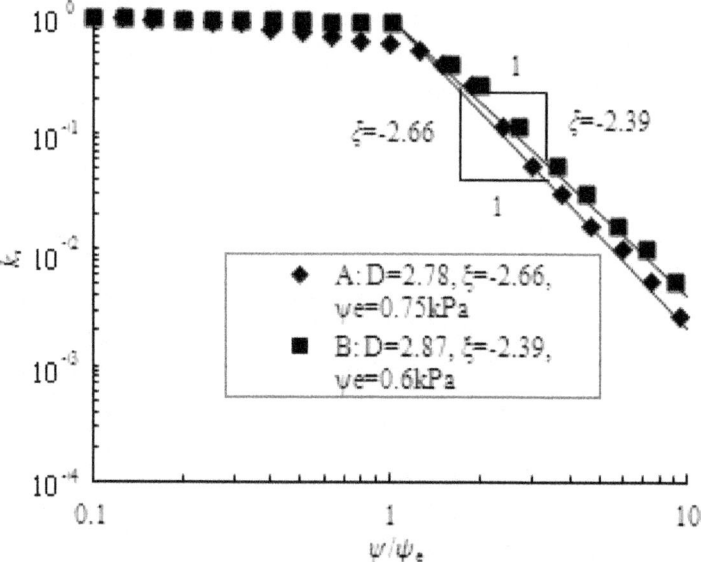

Figure 17. Comparison between the predictions of Eq. (44) and experimental data of the RHC (Data from Vogel and Roth, 1998).

Uno et al. (1998) studied the relationship between the pore-size distribution

(PSD) and unsaturated hydraulic properties of Toyoura sand. The relationship between the relative hydraulic conductivity (RHC) and the normollized volumetric water content Θ can be calculated from Eq. (45) using the parameters listed in Table 1. The prediction of Eq. (45) satisfactorily agrees with the experimental data of relative hydraulic conductivity (RHC) for Toyoura sand in Fig. 18. The proposed function of relative hydraulic conductivity (RHC) (Eq. (45)) is verified by the experimental data in Fig. 18.

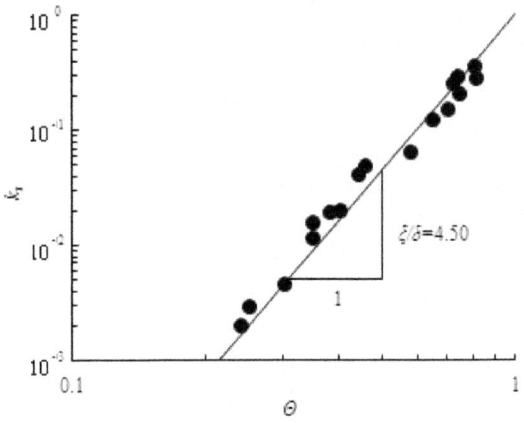

Figure 18. Comparison between predictions of Eq. (45) and experimental data of the RHC for Toyoura sand. (Data from Uno et al., 1998).

Prediction of RHC and SWD from SWCC

The characteristics of the pore-size distribution (PSD) were prerequisite to determine the relative hydraulic conductivity and soil-water diffusivity of unsaturated soils in Eqs. (44), (45), (47) and (48). A close relationship was established between the soil-water characteristic curve (SWCC) and the pore-size distribution by fractal model for the pore surface. The pore-size distribution (PSD) can be estimated from the soil-water characteristic curve because many soils show that the soil-water characteristic curve is equivalent to the pore-size distribution (PSD) function (Watabe et al., 2000). Thus, the fractal dimension of the pore surface and the air-entry value can be estimated from the fitting of the soil-water characteristic curve (SWCC). Thus, unsaturated hydraulic properties, such as relative hydraulic conductivity (RHC) and soil-water diffusivity (SWD) can be predicted using the fractal dimension and air-entry value obtained from the soil-water characteristic curve (SWCC).

Smettem and Kirkby (1990) offered the hydraulic properties of haploxeroll loam. Experimental data of the soil-water characteristic curve (SWCC) were shown in Fig. 19. The fractal dimension of the pore surface can be obtained from the fitted soil-water characteristic curve (SWCC). Through fitting the soil-water characteristic curve (SWCC), we obtained that δ=-0.37 for haploxeroll loam from Fig. 19. Fractal dimension of the pore surface can be calculated from Eq. (29) using the fitting value of δ. The fractal dimension (D) of the pore surface is 2.63 for haploxeroll loam. The air-entry value is 0.14kPa obtained from Fig. 19.

The hydraulic properties of unsaturated soils can be predicted using the fractal dimension of the pore-size distribution (PSD), obtained from fitted soil-water characteristic curve (SWCC). The fundamental parameters for predictions of hydraulic properties were tabulated in Table 3. The comparisons between predicted result and experimental data of the relative hydraulic conductivity (RHC) were shown in Fig. 20. The coefficient of hydraulic conductivity at saturation is 140 mmh^{-1}. Fig. 20 depicted the relationship between the relative hydraulic conductivity (RHC) and water potential for haploxeroll loam. The parameter ξ was -3.11 calculated from Eq. (44) using the fractal dimension D=2.63 for haploxeroll loam. The predictions of the relative hydraulic conductivity (RHC) calculated from the proposed function Eq. (44) are in satisfactory agreement with the experimental data.

The relationship between the soil-water diffusivity and water potential for haploxeroll loam was also shown in Fig. 20. The values of k_s and ψ_e are 140mmh^{-1} and 0.14kPa, respectively, used in Eq. (47). The parameters ς was -1.74, calculated from Eq. (47) using the fractal dimension D=2.63. The prediction of the soil-water diffusivity was the line passing through the point of (0.14, 0.64) with the slope of -1.74. The prediction of the soil-water diffusivity nearly agreed with the experimental data. The proposed function for the soil-water diffusivity was verified by the comparison between prediction and measured results. The parameters obtained from the soil-water characteristic curve (SWCC) and used to predict the relative hydraulic conductivity (RHC) and soil-water diffusivity (SWD) for haploxeroll loam were listed in Table 3.

Table 3. Parameters used to predict the unsaturated hydraulic conductivity.

Soil type	D	δ	ξ	ξ/δ	ς	$\psi_e(kPa)$
Haploxeroll loam	2.63	-0.37	-3.11	8.41	-1.74	0.14
Toyoura sand (wetting)	1.60	-1.40	-6.2	4.43	-3.8	/
Toyoura sand (drying)	1.26	-1.74	-7.22	4.15	-4.48	/
Loamy sand	2.68	-0.32	-2.96	9.25	-1.64	2.0

| McGee Ranch soil | 2.51(GSD) | -0.49 | -3.47 | 7.08 | -1.98 | 4.6 |

Figure 21 shows the experimental data and the fitting of the soil-water characteristic curve (SWCC) for loamy sand. The experimental data are scaled from Simunek et al. (1999). The parameter δ was -0.32, which is obtained from the fitting of the soil-water characteristic curve (SWCC) for loamy sand fromFig. 26. Hence the fractal dimension of the pore surface was 2.68 calculated from Eq. (29) using the fitting value of δ for loamy sand. The air-entry value was given by the intersection between the line expressed by Eq. (29) and the line of $\Theta=1$ in log-log plot. The air-entry value is 2kPa obtained from Fig. 21.

The parameters used to predict the hydraulic properties of loamy sand were listed in Table 3. Substituting D=2.68 and ψ_e=2kPa in Eq. (44), the predictions of relative hydraulic conductivity (RHC) can be obtained. The comparisons between predictions and experimental data of the relative hydraulic conductivity were shown in Fig. 22. The parameter ξ is -2.96 calculated from Eq. (44) using the fractal dimension D=2.68 for loamy sand. The predictions of the relative hydraulic conductivity were in good accord with the experimental data. The proposed function for the relative hydraulic conductivity was validated by the good agreement between the predictions and measurements.

The parameters of fractal model were first calibrated by matching the measured moisture-suction data. It should be noted that the experimental values of hydraulic conductivity were not matched by adjusting the model parameters. The calibrated SWCC model parameters were, instead, directly used to predict the relative hydraulic conductivity. The soil-water characteristic curves (SWCC) and their calibration of eight soils were shown in Fig. 23. The calibration of SWCC model parameters were listed in Table 4. As seen in these figures, the measured moisture-suction data for these soils were unavailable for the full range (0%-100%) of the degree of saturation. Predicting the suction beyond the available experimental data range was a challenging task since the pattern of variation was unknown (Ravichandran and Krishnapillai, 2011).

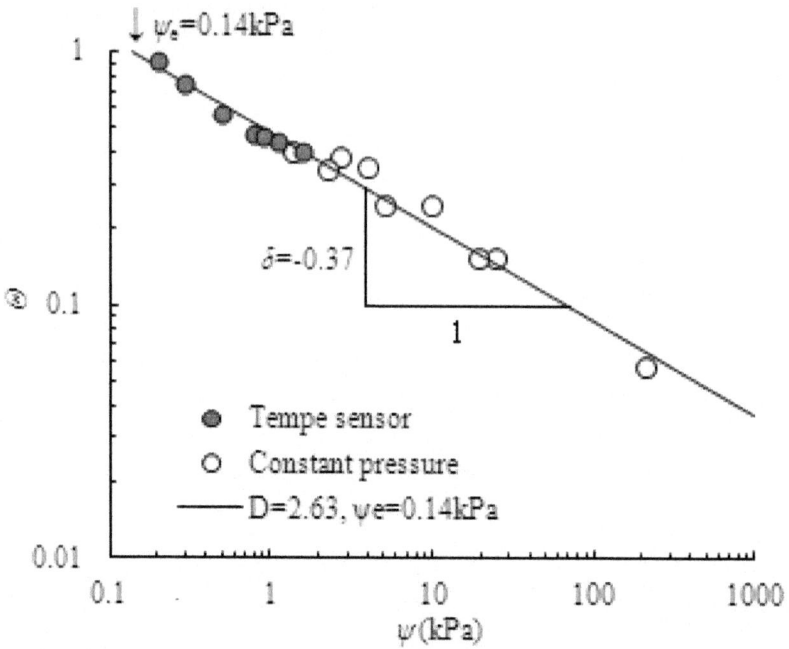

Figure 19. Fractal dimension of the pore surface and air-entry value evaluated from SWCC (Data from Smettem and Kirkby, 1990).

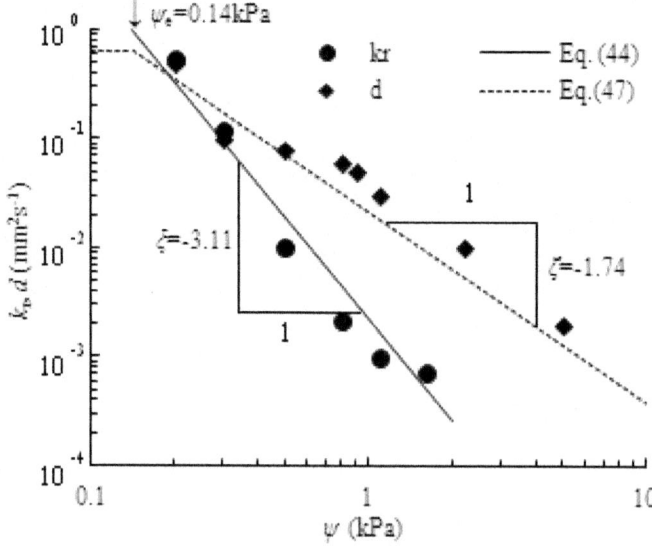

Figure 20. between predictions and experimental data of RHC and SWD for haploxe-roll loam (Smettem and Kirkby, 1990).

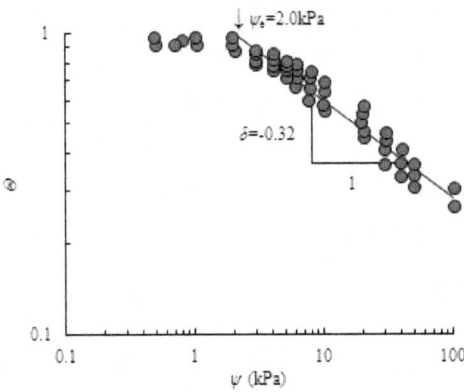

Figure 21. Fractal dimension and the air-entry value obtained from the fitting of SWCC for of loamy sand (Data from Simunek et al., 1999).

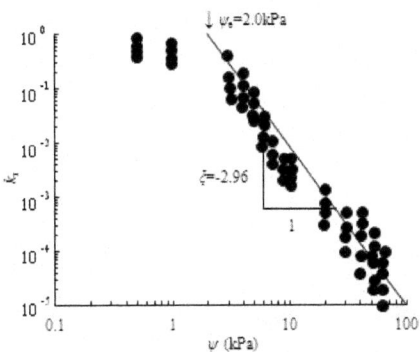

Figure 22. Comparison between measurements of RHC and prediction of Eq. (44) for loamy sand (Data fromSimunek et al., 1999).

Table 4. Parameters of the fractal model for eight selected soils.

Soil	$_\theta S$	$S_r(\%)$	D	$\psi_e(kPa)$	Reference
Lakeland	0.375	30-100	1.75	2.5	Elzeftawy and Cartwright, 1981
Superstition	0.5	30-100	2.7	1.2	Richards, 1952
Columbia sandy loam	0.458	50-100	2.29	4.5	Brooks and Corey, 1964
Touchet silt loam	0.43	20-100	2.09	7	Brooks and Corey, 1964
Silt loam	0.396	50-100	2.59	15	Reisenauer, 1963
Guelph loam	0.52	45-100	2.75	3	Elrick and Bowmann, 1964
Yolo light clay	0.375	45-100	2.87	1.5	Moore, 1939
Speswhite kaolin	0.56	55-100	2.85	15	Peroni et al., 2003

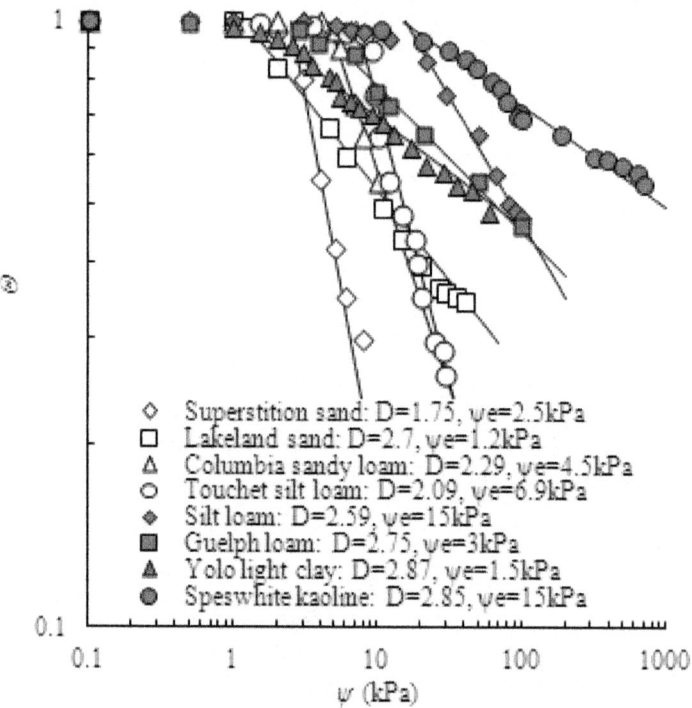

Figure 23. Fractal dimension and air-entry value obtained from the fitting of SWCC.

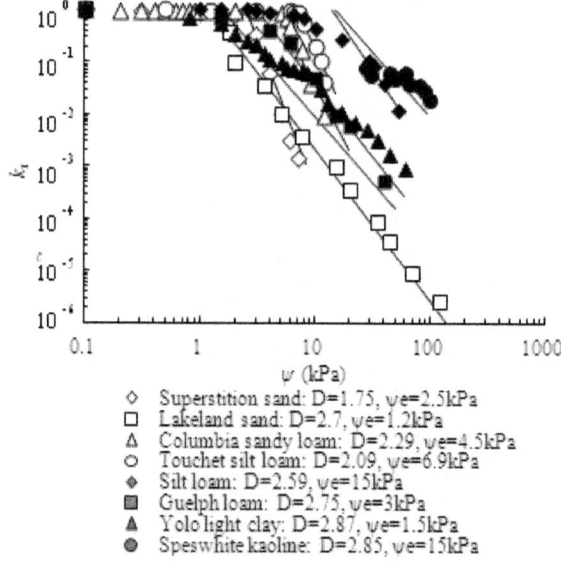

Figure 24. Comparison between predictions of Eq. (44) experiments of RHC.

The relative hydraulic conductivity of the above mentioned eight soils were predicted using the fractal model with the same fitting parameters listed in Table 4. The fitting parameters were obtained by matching the experimental SWCC. Fig. 24 illustrated the prediction of Eq. (44) and experimental data of relative hydraulic conductivity. It should be noted that the fractal model parameters were not calibrated or adjusted to match the measured hydraulic conductivity values. The predictions of Eq. (44)were comparable for sand and loam, as shown in Fig. 24. The fractal model showed slight deviations from the measured data of clay. In the case of Yolo light clay and Speswhite kaolin, the difference between the predictions and the experimental data increased as the suction increases (Fig. 24). The fractal model showed better predictions at lower suction range. However, the accuracy of fractal model in the higher suction range could not be verified because the experimental results were available only for the lower suction ranges.

It is seen that it is feasible to estimate the relative hydraulic conductivity of unsaturated soils using the fractal dimension of the soil pore surface from Figs. 19-24. A practical method to express the relative hydraulic conductivity was proposed to use the fractal dimension and the air-entry value, which can be obtained from the fitting of the soil-water characteristic curve (SWCC).

Prediction of RHC and SWD from GSD

The feasibility to determine the soil hydraulic properties using the fractal dimension of the grain-size distribution (GSD) is examined by the experimental data of McGee Ranch soil. The grain-size distribution (GSD) was measured by Hunt and Gee (2002) and was shown in Fig. 25. The fractal dimension of the grain-size distribution (GSD) is 2.51, obtained from Fig. 25.

Hunt and Gee (2002) gave the values of θ_s and θ_r were 0.4 and 0.01, respectively, the air-entry value was 4.6kPa. Using the fractal dimension of the grain-size distribution (GSD), the soil-water characteristic curve (SWCC) and the relative hydraulic conductivity (RHC) can be determined. The parameters of the grain-size distribution (GSD) used to determine the hydraulic properties are listed inTable 3. Comparisons between the predictions of Eq. (29) using the fractal dimension of the grain-size distribution (GSD) and the experimental data of soil-water characteristic curve (SWCC) are shown inFig. 26. It was seen that a good result is obtained to predict the soil-water characteristic curve (SWCC) using the fractal dimension of the grain-size distribution (GSD).

Hunt and Gee (2002) gave the values of saturated hydraulic conductivity was 0.001cm/s. Comparisons between the predictions of Eqs. (44) and (45) and experimental data were shown in Fig. 27. The prediction of Eqs. (44) and (45) nearly satisfied the experimental data of relative hydraulic conductivity in Fig.

27. In Fig.27, the predictions of Eq. (45) deviated from the experimental data near saturation, and extend to equal to the experimental results at low water content. The soil hydraulic conductivity can be approximately determined using the fractal dimension of the grain-size distribution (GSD).

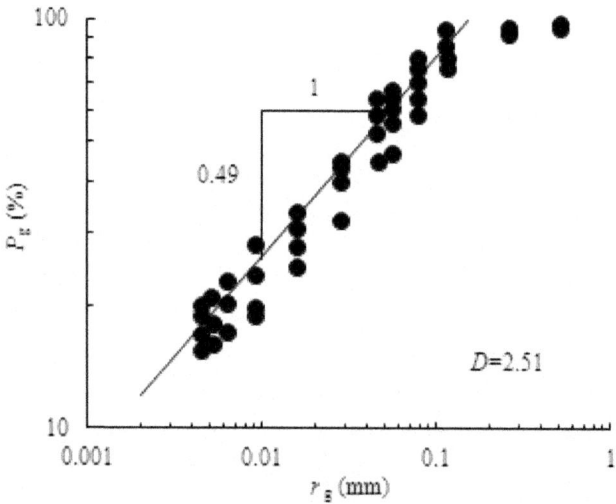

Figure 25. Fractal dimension of GSD for McGee Ranch soil (Data from Hunt and Gee, 2002).

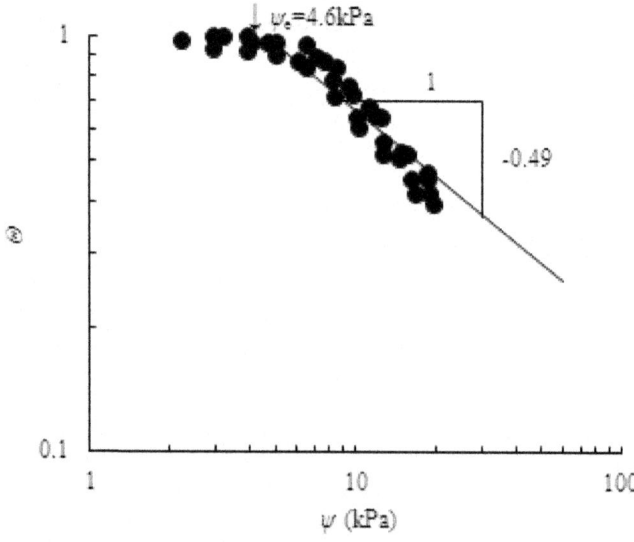

Figure 26. Comparisons between predictions and experiments of SWCC for McGee Ranch soil (Data from Hunt and Gee, 2002).

COMPARISON OF THE FRACTAL MODEL WITH THE VAN GENUCHTEN–MUALEM (G-M) MODEL

Van Genuchten–Mualem (G-M) Model For Relative Hydraulic Conductivity (RHC)

van Genuchten (1980) derived an empirical relationship to describe the soil-water characteristic curve (SWCC):

$$\Theta = \frac{1}{(1 + |\alpha\psi|^n)^m}$$

(49)

where α, n and m are the van Genuchten curve-fitting parameters, and m=1-1/n.

Burdine (1953) and Mualem (1976) presented a similar model to estimate the unsaturated hydraulic conductivity from pore size distributions inferred from soil water retention characteristics. The models share a number of similarities, allowing them to be written in a general form as (Hoffmann-Riem et al., 1999)

$$k_r = \frac{k_w}{k_s} = \Theta^l \left(\frac{\int_0^\Theta \psi^{-\chi}d\Theta}{\int_0^1 \psi^{-\chi}d\Theta} \right)^\gamma$$

(50)

where k_w is the unsaturated hydraulic conductivity at any water content, k_s is the saturated hydraulic conductivity, Θ is the normalized water content, ψ is matric suction. Generally accepted parameter values for the parameters (l, χ, γ) are (0.5, 1, 2) and (2, 2, 1) for the Mualem (1976) and the Burdine (1953) models, respectively.

Using the Mualem model, van Genuchten (1980) derived a closed-form to determine the relative hydraulic conductivity (RHC) at a degree of saturation,

$$k_r = \frac{\left[1 - |\alpha\psi|^{n-1}(1 + |\alpha\psi|^n)^{-m}\right]^2}{(1 + |\alpha\psi|^n)^m}$$

(51)

$$k_r = \Theta^{\frac{1}{2}} \left[1 - \left(1 - \Theta^{\frac{1}{m}}\right)^m\right]^2$$

(52)

Equations (51) and (52) are the representations of the G-M model, which is the most widely and popularly used to predict the hydraulic conductivity.

The other models for relative hydraulic conductivity (RHC) were also introduced as follows.

Van Genuchten–Burdine (G-B) Model For Relative Hydraulic Conductivity (RHC)

Hydraulic conductivity resulting from the Burdine capillary models (Burdine, 1953) was expressed as

$$k_r = \Theta^2 \left[1 - \left(1 - \Theta^{\frac{1}{m}} \right)^m \right]$$

(53)

Replacing Eq. (49) in Eq. (53) yielded

$$k_r = \frac{\left[1 - |\alpha\psi|^{n-2}(1 + |\alpha\psi|^n)^{-m} \right]}{(1 + |\alpha\psi|^n)^{2m}}$$

(54)

where m=1-2/n.

Note that for both Eqs. (52) and (54), because m is less than 1, the derivative dk/dθ is infinite for θ_s, whatever the value of n.

Van Genuchten–Fatt & Dykstra (G-Fd) Model For Relative Hydraulic Conductivity (RHC)

The capillary model proposed by Fatt and Dykstra (1951) was expressed as

$$k_r = \frac{\int_0^\Theta (\Theta - x)\psi(x)^{-2-c}dx}{\int_0^1 (1 - x)\psi(x)^{-2-c}dx}$$

(55)

Evaluating $\psi(x)$ from Eq. (49), the integrals of Eq. (55) with the condition of m>-(2+c)/2n and n>2+c resulted in:

$$k_r = \frac{\Theta B(a_1, b)I_{\Theta^{1/m}}(a_1, b) - B(a_2, b)I_{\Theta^{1/m}}(a_2, b)}{B(a_1, b) - B(a_2, b)}$$

(56)

where a_1=m+(2+c)/n, a_2=2m+(2+c)/n, b=1-(2+c)/n, I_x(a, b) and B(a, b) are the Incomplete Beta and Beta functions of positive arguments a and b, respectively, and given by:

$$B(a, b) = \int_0^1 x^{a-1}(1 - x)^{b-1}dx, \quad I_y(a, b) = \frac{\int_0^y x^{a-1}(1 - x)^{b-1}dx}{B(a, b)}$$

(57)

The results presented by Touma (2009) were obtained with c=0.5 and a_1=1. Even though the conductivity predicted by the capillary model of Fatt and

Dykstra (1951) gave the best results compared with both the quasi-analytical solution and observations, the resulting expression was not easy to use because it was necessary to evaluate Incomplete Beta and Beta functions. In order to simplify G-FD model, Θ was expressed by Eq. (49) and k_r was according to Eq. (31) with $\omega=2+2.5/mn$. The fitted result in a value of b was close to -mn, and a conductivity curve close to that predicted by Eq. (56).

Brooks & Corey-Brutsaert Model For Relative Hydraulic Conductivity (RHC)

Brutsaert (2000) presented a capillary model expressed as Eq. (37). Combining this capillary model with the soil-water characteristic curve (SWCC) of BC model (Brooks and Corey, 1966), relative hydraulic conductivity was written as Eq. (31) in form, and the parameter ω was given by

$$\omega = 2 - \frac{2}{b} \tag{58}$$

Combining of BC model for the soil-water characteristics with the Mualem and the Burdine capillary models, the relative hydraulic conductivity curves were also expressed as Eq. (31) in form, and the parameter ω was given by

$$\omega = 3 + \frac{2}{mm} \tag{59}$$

when BC is applied with the Burdine condition, and

$$\omega = 2.5 + \frac{2}{mm} \tag{60}$$

when applied with the Mualem condition.

Here we focused on the G-M model, as it was the most widely used in soil science and hydrology.

van Genuchten (1980) found that the predictions of the G–M model were less accurate for Beit Netofa clay. The fractal dimension and the air-entry value evaluated from the fitted soil-water characteristic curve (SWCC) are 2.89 and 60kPa, respectively (Fig. 28) The fitted parameters of the van Genuchten model, α and n are 0.0015 and 1.17, respectively from Fig.28. The parameters for the determination of the relative hydraulic conductivity (RHC) are listed in Table 5.

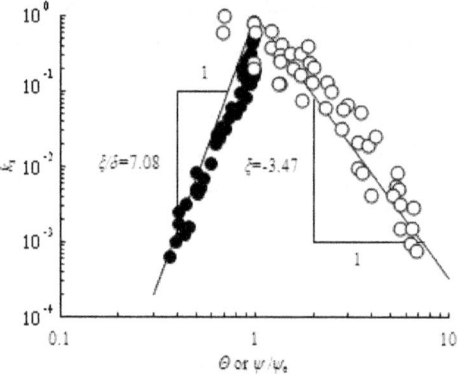

Figure 27. Comparisons between predictions and experimental data of RHC for Mc-Gee Ranch soil (Data fromHunt and Gee, 2002).

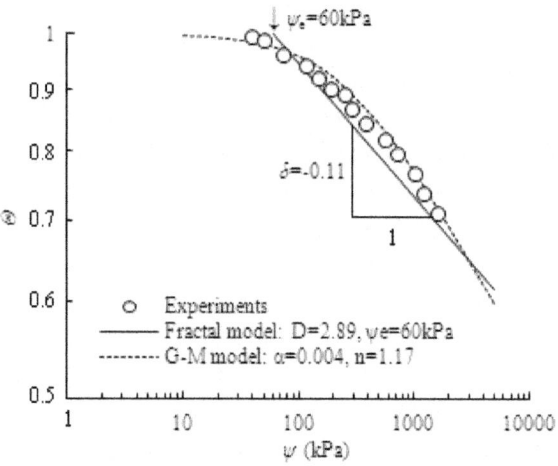

Figure 28. Fractal dimension and the parameters of G-M model obtained from SWCC for Beit Netofa clay (Data from van Genuchten, 1980)

Table 5. Parameters obatined from SWCC for prediction of RHC.

Soil type	Fractal model				G-M model	
	δ	D	ξ	ξ/δ	α (kPa^{-1})	n
Beit Netofa clay	-0.11	2.89	-2.33	21.2	0.004	1.17
Hanford glacial soil 0-099	-0.62	2.38	-3.86	6.23	0.015	1.71
Hanford glacial soil 2-1637	-0.82	2.18	-4.46	5.44	0.076	1.89
Grenoble sand	-0.64	2.36	-3.92	6.13	0.4	2.17

The saturated and residual water contents of Beit Netofa clay are 0.446 and 0, respectively. Comparisons between the predictions of both the fractal model (Eq. (44)) and the G–M model (Eq. (51)) and the experimental data of the relative hydraulic conductivity (RHC) for Beit Netofa clay are shown in Fig. 29. It is seen that predictions of the fractal model (Eq. (44)) satisfactorily agree with the experimental data, especially at high matric suction, while the predictions of the G–M model (Eq. (51)) are found to deviate from the experimental data. However, the fractal model shows a sharp corner at the air-entry value point, which is disagree with the measured data that usually show smooth transition after the air-entry value point. This default does not reduce the advantages to predict the relative hydraulic conductivity (RHC) using the fractal dimension obtained from the pore-size distribution (PSD), because the hydraulic conductivity near the point of the air-entry value nearly equal to the saturated conductivity.

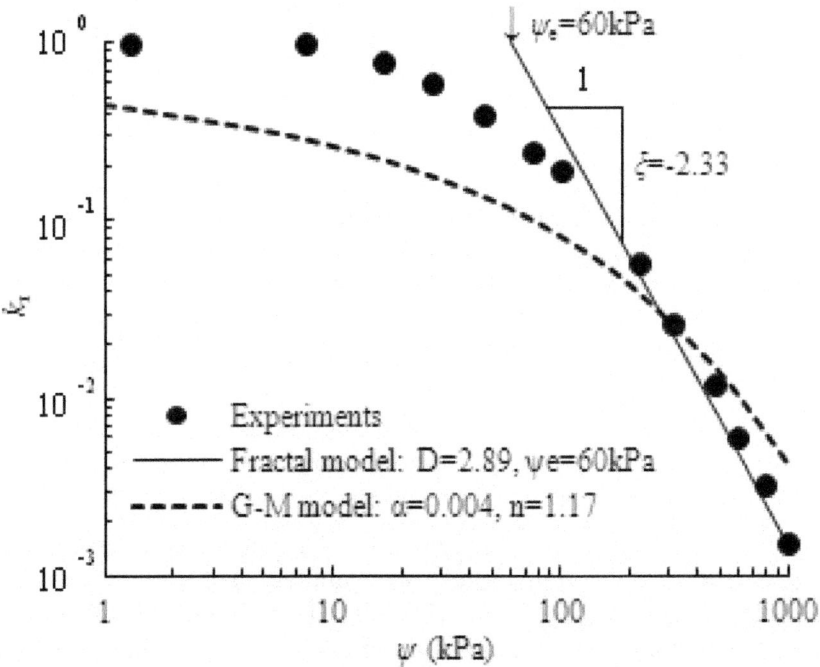

Figure 29. Comparison between the predictions of the RHC using fractal model and G-M model (van Genuchten, 1980).

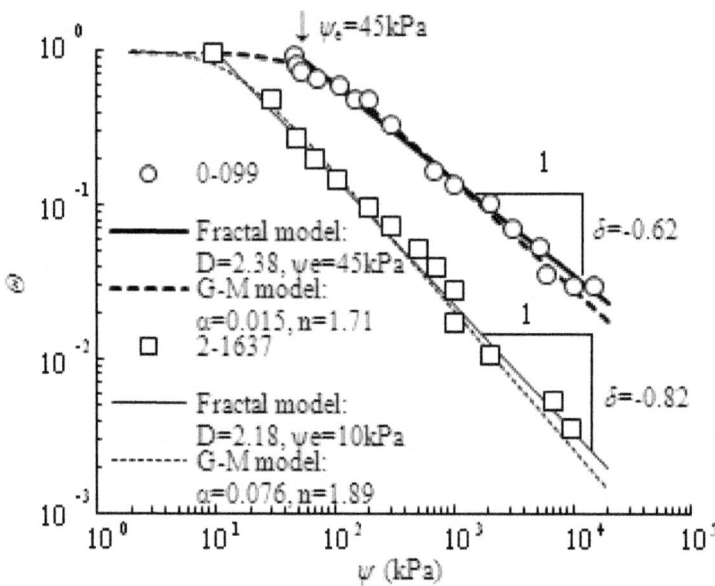

Figure 30. Fractal dimension and G-M parameters obtained from SWCC for Hanford glacial (Khaleel and Relyea, 1995).

The measured soil–water characteristic curves of Hanford glacial soil by Khaleel and Relyea (1995) were shown in Fig. 30. The saturated water contents (θ_s) are 0.338 and 0.303 for the samples 0–099 and 2–1637, respectively. The residual water contents (θ_r) are 0.039 and 0.025 for the samples 0–099 and 2–1637, respectively. From Fig. 30, it is seen that the soil-water characteristic curves (SWCC) of Hanford glacial soil are in good accord with the fractal model. The fractal dimensions (D), according to Eq. (29), are 2.35 and 2.18 obtained from the soil-water characteristic curves for the samples 0–099 and 2–1637 of Hanford glacial soil, respectively.

From the experimental data of the soil-water characteristic curve (SWCC), the parameters for determining relative hydraulic conductivity (RHC) using the fractal dimension are listed in Table 5. The larger the fractal dimension, the larger the complexity of the pore connectivity will be. The larger the fractal dimension, the larger the change in the water content with the same change in matric suction will occur. From the fittings of the soil–water characteristic curve in Fig. 30, the parameters (a and n) in the van Genuchten model (1980) for the soil–water characteristic curve (SWCC) are obtained, and are also listed in Table 5. Comparisons between the predictions of the fractal model (Eq. (45)) and the experimental data of the relative hydraulic conductivity (RHC) for Hanford glacial soil are shown inFig. 31. The predictions of the fractal model

(Eq. (45)) nearly agree with the experimental data of the relative hydraulic conductivity (RHC) for Hanford glacial soil in Fig. 31. The predictions of the G–M model (Eq. (52)) deviate from the experimental data of the relative hydraulic conductivity (RHC) inFig. 31. It is seen that the predictions of the fractal model are better than those of the G–M model for Hanford glacial soil in Fig. 31.

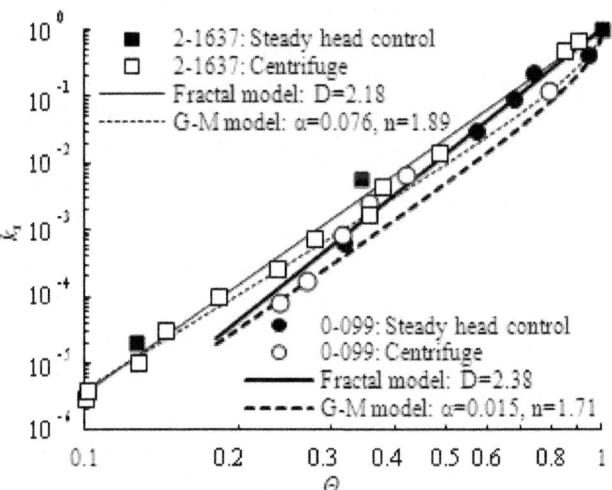

Figure 31. Comparison between the predictions using fractal model and G-M model (Data from Khaleel and Relyea, 1995).

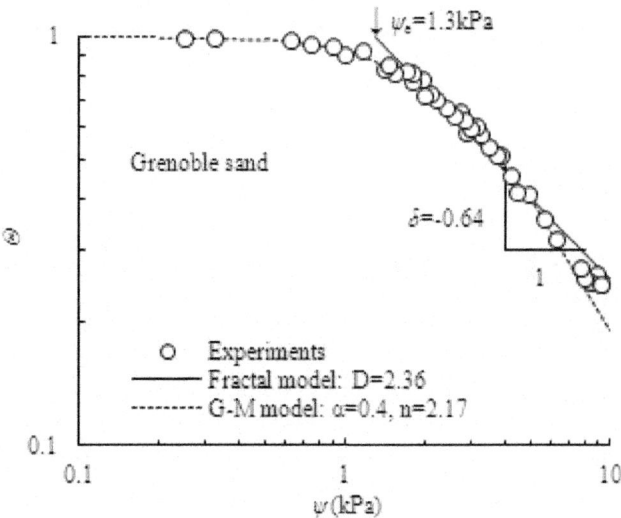

Figure 32. Fitted SWCC curves.

The combinations of Eq. (29) with fractal capillary model were tested on Grenoble sand. Data points for the soil-water characteristic curve (SWCC) and hydraulic conductivity were taken from Touma (2009). The infiltration experiment was conducted under a constant head of 2.3cm (Touma & Vauclin, 1986). The solid lines in Figure 37 were the fitted curves for Eq. (29), and the resulting predicted hydraulic conductivity were shown in Fig. 32. Fractal model gave good results for the Grenoble sand. The parameters obtained through the fitting results of soil-water characteristic curves (SWCC) were used to predict hydraulic conductivity curve. Comparison between the fractal model and the G-M model were shown in Figs. 33. It was found that the prediction of hydraulic conductivity curve using Eq. (45) closed to the experiments for Grenoble sand.

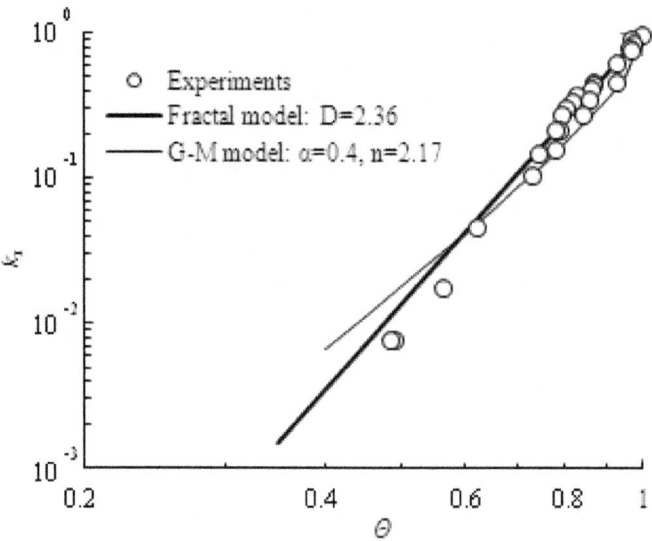

Figure 33. Comparison between fractal model and G-M model.

RAINFALL INFILTRATION OF UNSATURATED SOILS

Rainfall induced slope failures often occur as relatively shallow failure surfaces orientated parallel to the slope surface and are observed and analyzed by different mechanisms. The effect of seepage on slope stability is introduced in the analyses by calculating the critical depth for an infinite slope with rainfall infiltration, while the effect of both negative and positive pore water pressures on the stabilities of initially unsaturated slopes are explained and coupled with infinite slope analysis and pore-air flow analysis methods in order to present a predictive formulation of slope failures that occurs in rainfall evens and is derived from a fractal model on unsaturated soil. The formulation serves as a baseline analysis method for evaluating potentially unstable slopes.

The pore water pressure pattern that develops in the initially unsaturated soil will occur as a transient process as the infiltration moves downward into the soil profile. Several factors should been taken into account in unsaturated analyses. The shear strength of the soil mass and the development of seepage forces which both depend on the evolution of the pore water pressure profile must been addressed in detail.

Here, an individual soil slice can be treated as a one-dimensional column with vertical infiltration. Suppose that any lateral flow between the adjacent slices will be equal on the up-slope and down-slope boundaries. Then considering vertical infiltration only could meet the flow continuity requirement. These one-dimensional infiltration analyses are addressed by the saturated/unsaturated seepage finite element software SEEP/W (GEO-SLOPE International Ltd.2007) coupled with the air flow software AIR/W (GEO-SLOPE International Ltd.2007).

For a homogenous, isotropic soil, one-dimensional water flow equation (GEO-SLOPE International Ltd.2007) is given by:

$$m_w \gamma_w \frac{\partial H_w}{\partial t} = \frac{\partial}{\partial y}\left(k_w \frac{\partial H_w}{\partial t}\right) + m_w \frac{\partial P_a}{\partial t} + Q_w$$

(61)

where H_w is the total water energy potential comprised of both pressure and elevation potentials; P_a is the pore air pressure; kw is hydraulic conductivity; y is y coordinate; m_w is the slope of SWCC; t is time; and Q_w has units of length per time.

For the air conversation of mass, we can arrive at the pore air general mass balance equation (GEO-SLOPE International Ltd.2007) as like:

$$\left(\rho_w m_w + \frac{\theta_a}{\overline{R}T}\right)\frac{\partial P_a}{\partial t} = \frac{\partial}{\partial y}\left[\frac{\rho_a k_a}{\gamma_{0a}}\frac{\partial P_a}{\partial y} + \frac{\rho_a{}^2 k_a}{\rho_{0a}}\right] + \rho_a \lambda_w m_w \frac{\partial H_w}{\partial t}$$

(62)

where k_a is air permeability; θ_a is volumetric air content; ρ_a is air density; γ_{0a} is initial air unit weight; ρ_{0a} is initial air density; $\overline{R}$ is ideal air constant and $\overline{R}$ =287J/(kg K) for dry air; T is temperature; H_w is total head and P_a is pore air pressure. These two governing equations can be used to determine total pressure head profile for the water and the pore air pressure, and then suction is attained.

The relationships relating volumetric water content and hydraulic conductivity to suction must be known to solve the Eq. (61). And similarly to calculate the pore air pressure the pore air permeability function is also wanted. To find the effect of different fractal dimensions and air-entry values of different soil types, series of parameter studies have conducted. Different types of Soil-water characteristic curves and hydraulic conductivity curves were shown in Figs. 34 and 35. In the legends of these pictures, Soil 10, 2.1,-5, for example, means a type of soil with air-entry value, ψ_e=10kPa, the fractal dimension, D=2.1 and saturated hydraulic conductivity k_s=1×10^{-5}m/s. It is assumed that θ_r=0. The pore air permeability function curve used in these studies was shown in Fig.36, and it remained a constant (GEO-SLOPE International Ltd.2007).

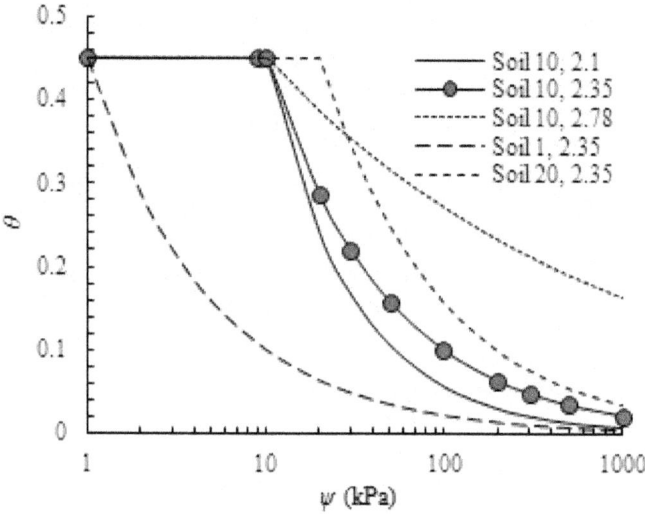

Figure 34. Soil-water characteristic curves for soils.

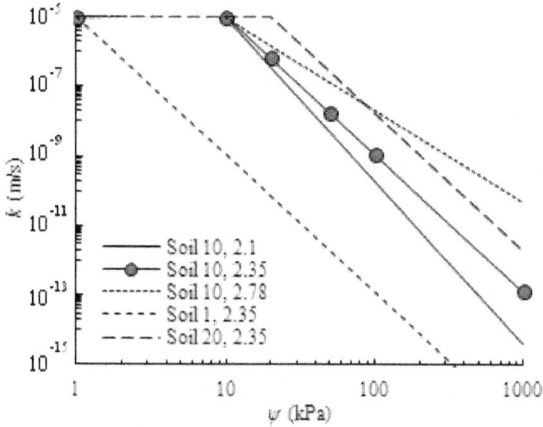

Figure 35. Hydraulic conductivity curves f.

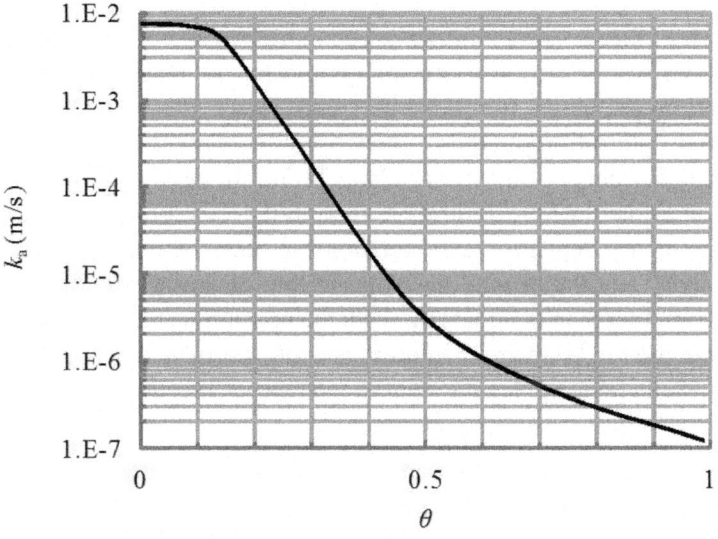

Figure 36. Pore air permeability function used in study.

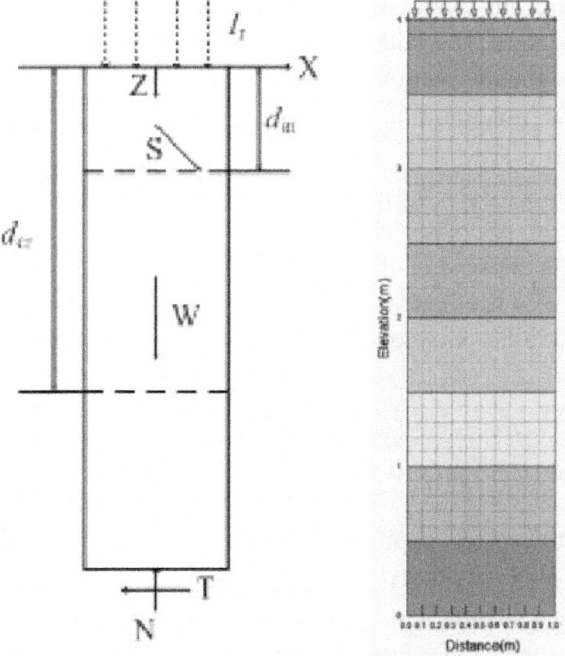

Figure 37. Sketch map of the one-dimensional infiltration model.

A 4m-deep soil column was modeled for the one-dimensional analyses, as shown in Fig. 37, and infiltration occurs from the top of the column, while the bottom was set to be a pervious boundary. A linear hydrostatic suction distribution presented for the initial conditions. Analyses for the rainfall infiltration into different types of soils were performed using the unsaturated characteristic curves shown in Figs. 34-36. In applying a top boundary condition to simulate the infiltration of rainfall, it is important to realize that an influx boundary depends on the relationship between the saturated hydraulic conductivity and the rainfall intensity. In the following analyses, if a rainfall intensity greater than or equal to k_s with the non-infiltrating rainfall running off the slope, the top boundary condition is set to a total head equal to 4m; while if field results indicate that the rainfall intensity is less than k_s, then a flux type boundary condition is more appropriate.

Since initial failures often have small depth-to-length ratios and form failure planes parallel to the slope surface, the use of infinite slope analysis in modeling the infiltration process by vertical one-dimensional analysis makes it justified in describing the physical process of failure initiation. However, the methods used in traditional infinite slope analysis must be modified to take into account the variation of the suction profile that results from the infiltration process. There are two distinct failure mechanisms can be initiated by the infiltration process. The failure takes place due to positive pore pressure and it takes place while the suction still exists.

The shear strength of unsaturated soil can be represented by fractal model, and written as follows (Xu, 2004):

$$\tau = c' + (\sigma - u_a) \tan \phi' + \psi_e^{3-D} \psi^{D-2} \tan \phi' \tag{63}$$

where τ is the shear strength; c' is the effective cohesion; $(\sigma-u_a)$ is the total normal stress; φ› is effective friction angle; ψ is the suction; ψ_e is the air-entry value. The stability envelope can be represented as (Collins and Znidarcic, 2004):

$$d_{cr} = \frac{c' + \gamma_w h_c \tan \phi' - \gamma_w h_p \tan \phi'}{\gamma \cos^2 \beta (\tan \beta - \tan \phi')} \tag{64}$$

where d_{cr} is the critical depth; β is slope angle; h_p is the pore passive pressure and h_c is the pore negative pressure.

We can get suction with analyses of both pore water pressure and pore air pressure in the soil slope under rainfall infiltration. Combining Eq. (63) with Eq. (64), the new stability envelope can be derived and written as follows:

$$d_{cr} = \frac{c' + \psi_e{}^{3-D}\psi^{D-2}\tan\phi' - \gamma_w h_p \tan\phi'}{\gamma\cos^2\beta(\tan\beta - \tan\phi')}$$

(65)

The critical depth for infinite slope failure is now a function of the suction, ψ, pressure head h_p, the given material and slope characteristics, c', ϕ>, γ, ψ_e, ξ, γ_w, and β. In this way, the slope stability issues related to the decrease in shear strength from a loss of soil suction and the development of seepage forces from positive pressure head generation can be clearly understood.

Because the infiltration results and the slope stability results were both presented in terms of the pressure head and the suction profile, the two analyses can be coupled to yield a comprehensive method for determining the location and time of failure for a slope if the soil, slope, and rainfall parameters are given. The methodology of the coupled analysis involves plotting a stability envelope as defined by Eq. (65) for specified soil and slope parameters, over a given infiltration profile generated from the particular rainfall and unsaturated characteristic curves. Coulomb failure and the initiation of slope mobilization are defined at the points where the infiltration trace intersects the stability envelope. Intersections of the infiltration trace and the stability envelope indicate points at which the slope is unstable and can be thought of as "critical depths" of failure. It must therefore be assumed that if the slope is initially stable, the point at which the stability envelope inter sects the initial suction distribution line will not contribute any further information about the failure mechanisms from infiltration.

Here, factors controlling the unsaturated soil slope under infiltration are studied and the results which combine infiltration analyses and stability analyses are shown in Figs.43-47. Infiltration analyses were performed by software SEEP/W coupled with AIR/W in order to calculate the suction file and the stability envelope as defined by Eq.(65) for slope parameters, β=40°, γ=19.6kN/m³, γ_w=10kN/m³, and shear strength parameters c'=5kPa, ϕ>=15°. Other parameters are studied in a systematic way.

Effect of Fractal Dimension

Soil column was saturated due to rainfall infiltration from the initial hydrostatic suction profile. It was shown that infiltration took place in unsaturated soil much faster for the soil with higher fractal dimension. Three fractal dimensions (D=2.1, 2.35 and 2.78) were calculated respectively and the results were shown in Fig.38. It was seen from Fig. 38 that the higher initial hydraulic conductivity existed on the top of soil column which had a higher fractal dimension. On the other hand, the stability envelope for the soil with higher fractal dimension was steeper. There was an obvious inflection point in the stability envelope of the soil with the low fractal dimension. The results also indicated that it was the relative shape of the unsaturated characteristic curves (Fig.34 and Fig. 35) that has a controlling effect on the suction distribution.

Effect of Air-Entry Value

Three air-entry values (ψ_e=1kPa, 10kPa and 20kPa) are calculated respectively and the results as shown in Figs.39. Compared with the results of three different air-entry values, both the infiltration trace and the stability envelope have changed with the different air-entry values. When the air-entry value is relatively small (ψ_e=1kPa), as shown in Fig.39a, passive pore water pressure was generated on the upper soil column at the time t=32h.

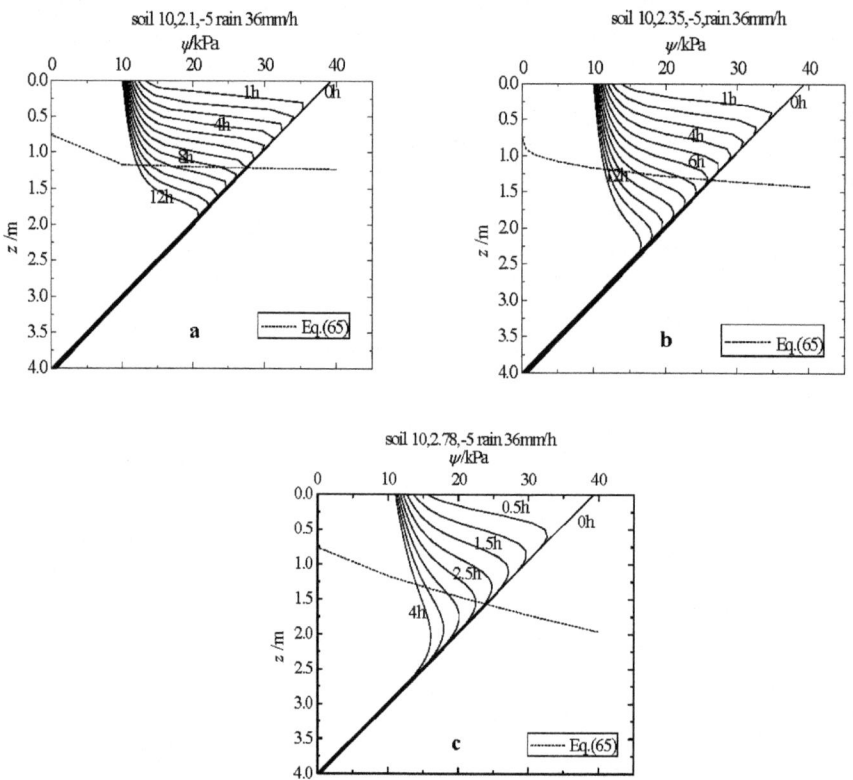

Figure 38. Infiltration results for soils with different fractal dimension and the new stability envelope.

Due to the low initial permeability on the top of soil column and a state of lower saturation, fewer pores initially filled with water and there are fewer channels available for fluid transport and consequently the flow of water is hindered. With the commencement of infiltration at the top boundary, water is forced into a soil which is not capable of transporting it efficiently, which results in the development of positive pressure heads as the infiltration front progresses downward. While the air-entry value is relatively larger, as shown in Fig.39c, infiltration moves downward much faster and suction has reached to the air-entry value. Passive pore water pressure occurs at the bottom of the soil column which means that there is little unsaturated zone in the middle of soil column during the rainfall infiltration. Besides, the stability envelope is much smooth when the soil has a relatively low air-entry value, and consequently the critical depth is smaller.

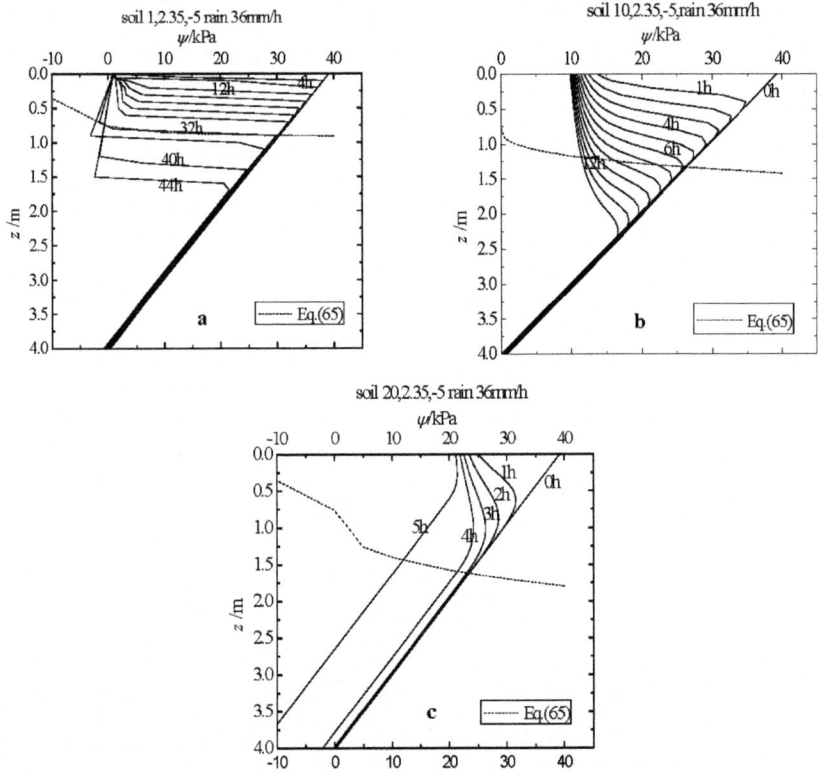

Figure 39. Infiltration results for soils with different air-entry value and the new stability envelope.

Effect of Saturated Hydraulic Conductivity

The dominate factors which control the stability of unsaturated soil slope under rainfall infiltration are the soil saturated hydraulic conductivity and rainfall intensity. While the soil saturated hydraulic conductivity determines the water transportation ability of soil and the infiltration quantity most depends on the rainfall intensity. In this parameter study series, three different saturated hydraulic conductivity values ($k_s = 1 \times 10^{-5}$m/s, 1×10^{-6}m/s, 1×10^{-7}m/s) are taken into account and the results are shown in Fig.40.

When the rainfall intensity is much larger than the saturated hydraulic conductivity, soil in the top region of the column has been saturated soon after the infiltration begins and its suction almost reaches to zero at the time t=1.5h. It can be seen that the suction at the base is increasing over time as a positive number, which means that the soil is actually de-saturating at its base while wetting up above. This is caused by the air which becomes trapped when the top has saturated and starts to resist water infiltration. As the situation that the rainfall intensity is much smaller than the saturated hydraulic conductivity as shown in Fig.40a, the suction of the top region decreases with the infiltration developing, however, the top region has remained unsaturated and the infiltration depth is also small due to the small rainfall quantity into the soil.

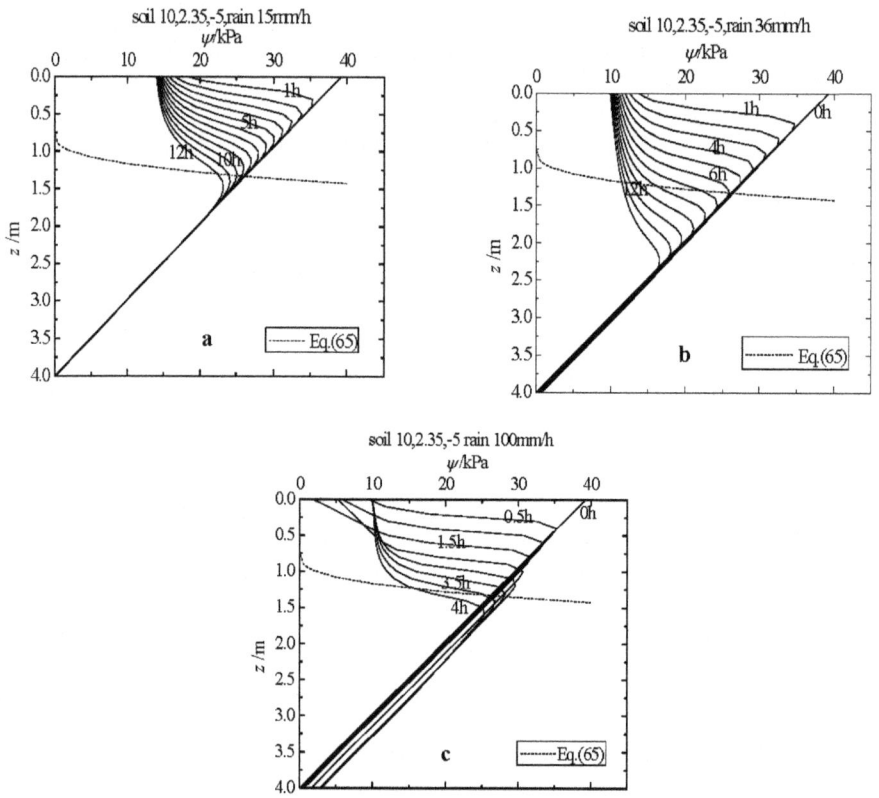

Figure 40. Infiltration results for soils due to different rainfall intensities.

Effect of Rainfall Intensity

The rainfall intensity for a certain saturated hydraulic conductivity has taken into considered to simulate the infiltration process. The range is changed from 0.3mm/h to 100mm/h and the relationship between the time of failure and rainfall intensity for the three soil types of saturated hydraulic conductivity has plotted in Fig.41. These curves decrease as exponent forms similarly while the magnitude of different curves has changed dramatically. The time at which the soil slope failure occurs can be about ten hours for the soil saturated hydraulic conductivity k_s equals to 1×10^{-5}m/s, and for the situation that k_s equals to 1×10^{-6}m/s, the time can range from 10 to 100 hours. While much more time is needed for the lower saturated hydraulic conductivity soil type.

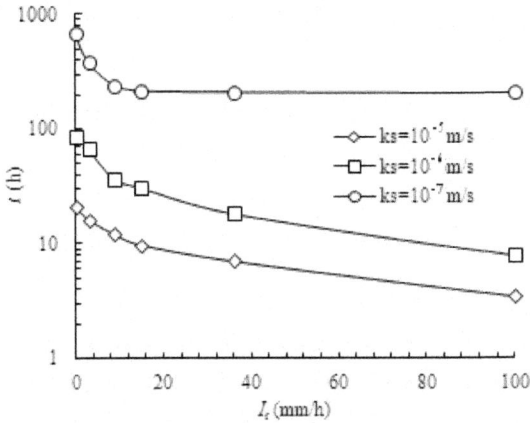

Figure 41. The time of failure vs. rainfall intensity.

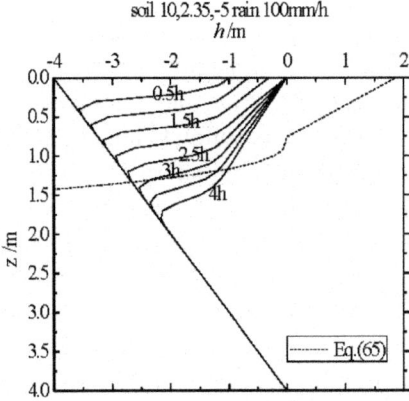

Figure 42. Infiltration results without air phrase consideration.

Infiltration Analyses Coupled With the Air Flow Analyses

In the infiltration analyses, air flow movement in the soil pore is also conducted. The effect of the pore air flow movement is especially obvious in the situation when the rainfall intensity is much larger than the soil saturated hydraulic conductivity. Compared with the results of infiltration analyses coupled with air flow analyses, the results without air phrase consideration are shown in Fig.42. In this condition, the negative pore water pressure gradually reaches zero at the top of soil column. While the pore water pressure remains unchanged at the bottom of the column. If we take the pore air flow movement into consideration, the results are different as shown in Fig.40c. Besides, the suction could be calculated more precisely when we attain both the pore water pressure and the pore air pressure by simulating infiltration analyses coupled with air flow analyses.

SLOPE STABILITY ANALYSES DUE TO RAINFALL INFILTRATION

It was widely recognized that the rainfall infiltration took a great role in causing landslides, while the relative importance of soil properties, rainfall intensity, initial water table depth and slope geometry in inducing instability of a homogenous unsaturated soil slope under different rainfall was investigated through a series of parametric studies (Rahardjo, et al, 2007). Soil properties and rainfall intensity were found to be the primary factors controlling the instability of slopes due to rainfall, while the initial water table depth and slope geometry only played a secondary role.

The factors affecting the stability of a slope were considered to be the soil properties, rainfall intensity, initial depth of the groundwater table, and the slope geometry (i.e., slope angle and slope height). To assess the effects and relative contribution of controlling factors, a series of parametric studies were performed on a typical geometry of a homogeneous soil slope shown in Fig. 43. It had the boundary conditions as follows: ab, bc, cd=q=I_r(rainfall intensity); ah, de, fg=q=$0 m^3/s$ (i.e. no flow boundary); and ef, gh=h_t(total head at the side). Four slope heights, Hs,(3m, 4m, 5m and 6m), three slope angles, α,(26.6, 33.7 and 45.0), eight initial depths of groundwater table (GWT), Hw,(0.5m, 1m, 2m, 3m, 5m, 7.5m, 10m and 15m), three fractal dimensions D (2.1, 2.35 and 2.78), four air-entry values ψ_e (1kPa, 10kPa, 20kPa and 50kPa), five values of saturated hydraulic conductivity, k_s (10^{-4}m/s, 10^{-5}m/s, 10^{-6}m/s, 10^{-7}m/s and 10^{-8}m/s) and five rainfall intensities Ir(3mm/h, 9mm/h, 15mm/h, 36mm/h and 100mm/h each for 24h duration) were used in six cases of parametric studies. The shear strength parameters of the soils used in the parametric study were c'=10kPa, effective angle of internal friction,φ>=26°, and unit weight of soil, γ=20kN/m³.

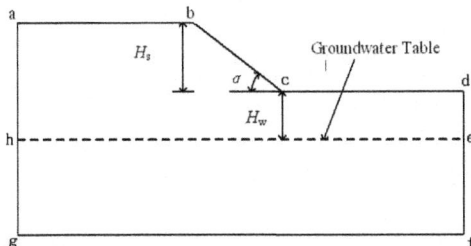

Figure 43. Sketch map of the slope in study.

The fractal model of the soil-water characteristic curves for unsaturated soil was written as Eq. (29). The function of relative hydraulic conductivity was used as Eqs. (44) and (45). The derived SWCC and RHC for the all six soils were shown in Figs. 34 and 35. In the legends of these pictures, the symbol S in the soil names represent "soil", the first number means the air-entry value, and the second one means the fractal dimension.

In the seepage analysis, the governing partial differential equation (GEO-SLOPE International Ltd.2007) for a two-dimensional transient water flow used in the finite element seepage model was written as follows:

$$m_w \gamma_w \frac{\partial (H - y)}{\partial t} = \frac{\partial}{\partial y}\left(k_w \frac{\partial H_w}{\partial t}\right) + m_w \frac{\partial P_a}{\partial t} + Q$$

(66)

where H is the total head; kx are ky ars hydraulic conductivity in x, y direction respectively ; Q is the applied boundary flux; m_w is the slope of the SWCC and γ_w is the unit weight of water. Eq. (66) was solved by SEEP/W software (Geo-Slope 2007). The boundary conditions used in the transient seepage analysis were shown in Fig. 43. And the initial condition for the analyses was a hydrostatic condition with a limiting pore water pressure of -75kPa. Then the SEEP/W software generated the negative pore water pressure as the initial condition. The pore-water pressures obtained from the seepage analysis were then used in the slope stability analyses to calculate the factor of the slope safety, F_s.

The equation for unsaturated shear strength was used as Eq. (63). In the study, we chose c'=10kPa,φ'=26°. The Morgenstern-Price Method was adopted in the slope stability analysis. This method considered both shear and normal inter-slice forces, satisfied both moment and force equilibrium, and allowed for a variety of user-selected inter-slice force function. The slope stability analysis using Morgenstern-Price Method was performed using SLOPE/W software

(Geo-Slope 2007). The pore-water pressures, u_w, obtained from the transient seepage analyses using SEEP/W were added to SLOPE/W to be incorporated in the slope stability analyses.

The results were presented with attention on the effects of the factors which impacting on the soil slope stability under rainfall for 24h with a combination of various controlling factors.

Effect of Soil Properties

The variation in factor of safety with time for a homogeneous soil slope of constant slope heightH_s=6m, initial groundwater table depth H_w=2m, subjected to rainfall intensities of 3, 9 and 15mm/h of respective soil for 24h with a combination of various soil types (different fractal dimensions and air-entry values) were analysized as follows.

Effect of fractal dimension

In Figs.44, the fractal dimension, D was 2, 2.35 and 2.78, respectively. The air-entry value, ψ_e remained 10kPa. It is shown that the reduction of slope stability is larger in the condition of larger fractal dimension under the condition of the same rainfall intensities and the saturated hydraulic conductivity. This means that the soil slope is much safe if the fractal dimension is low while the saturated hydraulic conductivity remains constant. The factor of safety decreases in a short time under rainfall infiltration when the fractal dimension is relatively large, as shown in Fig.44c, while the decrease occurs at a longer time for the low fractal dimension soil type, as shown in Fig.44a.

If the air-entry value and the saturated hydraulic conductivity remain unchanged, the slope of the hydraulic conductivity function is small when the fractal dimension is large. Then the initial hydraulic conductivity value at the beginning of the rainfall infiltration is much big, which results in the speed of rainfall infiltration is great and the saturated degree of soil inside of slope changes quickly leading unsaturated soil to become saturated. The passive pore water pressure generated by the increase of degree of soil saturation makes a negative impact on the instability of soil slope under rainfall.

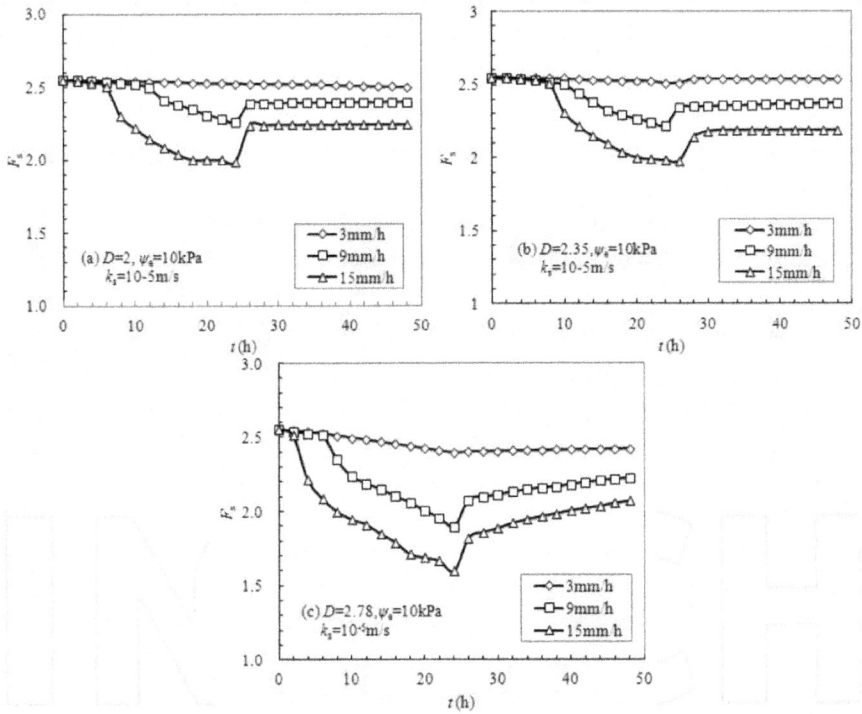

Figure 44. Effect of fractal dimension on slope stability.

Effect of air-entry value

The effects of air-entry value on the stability of soil slope due to rainfall infiltration are shown in Figs. 45. The air-entry value, ψ_e is 1kPa, 10kPa, 20kPa and 50kPa respectively and the fractal dimension, Dremains 2.35. The style of the slope safety curve changes a lot with different air-entry values in Figs. 45. In Fig. 45a, the air-entry value, ψ_e is 1kPa and the decrease of factor of safety occurs during 16~20h after rainfall began. As the air-entry value, ψ_e is 10kPa, shown in Fig. 45b, the decrease of slope safety factor occurs at 8h after rainfall began. At the beginning of rainfall infiltration, the decrease happens inFigs. 45c-d, in which the air-entry value is 20kPa and 50kPa respectively. This means that a delay time appears which is depended with the air-entry value.

And it is obvious that small air-entry value decides a relative long delay time. Besides, the factor of slope safety remains minimum during another 24 hours after rainfall stops in Fig. 45a, which has a low air-entry value (ψ_e=1kPa). While ψ_e=50kPa, the recovery of safety factor curve in Fig. 45c occurs and dramatically. This is also due to the style of hydraulic conductivity function, as shown in Fig. 35. The initial hydraulic conductivity value becomes large when the air-entry value increases with fractal dimension, saturated hydraulic conductivity and saturation degree remain the same.

Effect of saturated hydraulic conductivity

The effects of saturated hydraulic conductivity on stability of a homogenous soil slope are reflected through the relationship between F_s and k_s, shown in Fig. 46. The saturated hydraulic conductivity, k_s, was varied with five different values of 10^{-8}m/s, 10^{-7}m/s, 10^{-6}m/s, 10^{-5}m/s and 10^{-4}m/s for a homogeneous soil slope of constant soil type S10, 2.35 (ψ_e=10kPa, D=2.35), H_w=2m, H_s=6m, α=33.7°, subjected to rainfall for 24h with six rainfall intensities of 3, 9, 15, 36, 100 and 360mm/h. All the plots in Fig. 47 shows that soil with k_s values of 10^{-8}m/s and 10^{-7}m/s, respectively, are less affected by rainfall. Contrarily, soil with k_s values of 10^{-5}m/s and 10^{-4}m/s are greatly affected by rainfall. It is suggested that soil slopes with a low saturated hydraulic conductivity are relatively safe under short-duration rainfall infiltration and for the soil slopes with a high saturated hydraulic conductivity the stability is affected by the short-duration rainfall greatly. This means that slopes with low saturated hydraulic conductivity need a long-duration rainfall to intrigue the instability.

The F_s(min)-k_s critical curve was presented in the broken line in Fig. 47 as follows:

$$F_s(\text{min}) = \frac{a}{1 + be^{-ck_s}}$$

(67)

where F_s(min) is minimum factor of safety; k_s is saturated hydraulic conductivity; a, b and c are the fitting parameters and e is natural number (i.e., 2.71...). Here, a=1.293, b=-0.492, c=1.2×10^5, and the corresponding coefficient of correlation, r^2 is 0.9998.

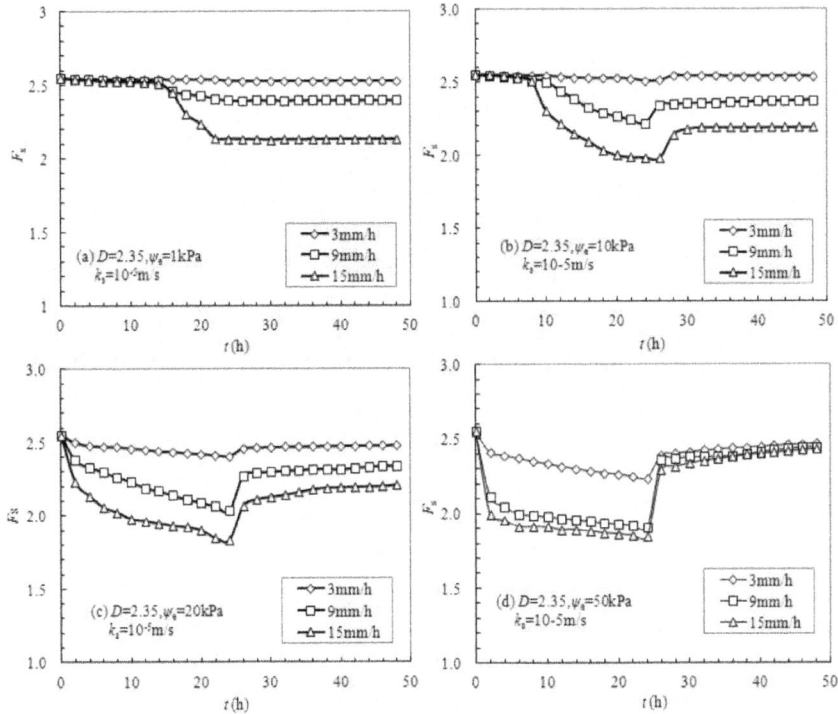

Figure 45. Effect of air-entry value on the slope stability.

Effect of rainfall intensity

The relationship between minimum factor of safety, F_s(min), versus logarithmic of rainfall intensity, I_r are plotted in Fig. 47. The semi log plot shows that generally the F_s(min) and I_r relationships follow a sigmoid shape in Fig. 47. There are two inflect points for the sigmoid shape line. The Fs(min) is almost constant at very low rainfall intensities for all soil types. And it starts to decrease rapidly when the first inflection point is reached. The trend in F_s(min) versus I_r relationship observed in Fig. 47 can be described by a sigmoid equation (MMF Line) as the form of

$$F_s(\min) = \frac{ab + cI_r{}^d}{b + I_r{}^d}$$

(68)

where F_s(min) is minimum factor of safety; I_r is rainfall intensity; and a, b, c, d are fitting parameters. The values for the fitting parameters and the corresponding coefficient of correlation, r^2, for the sigmoid line in Fig.47.

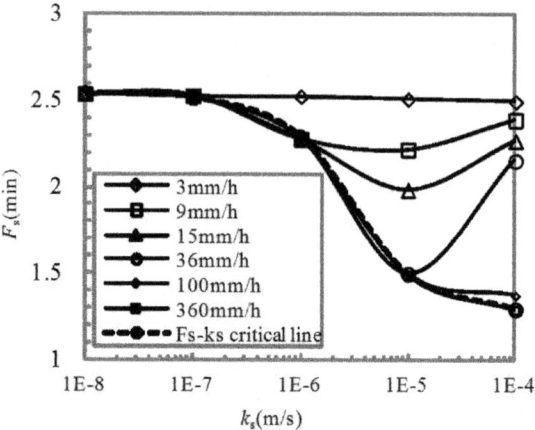

Figure 46. Relationship between K_s and minimum factor of safety.

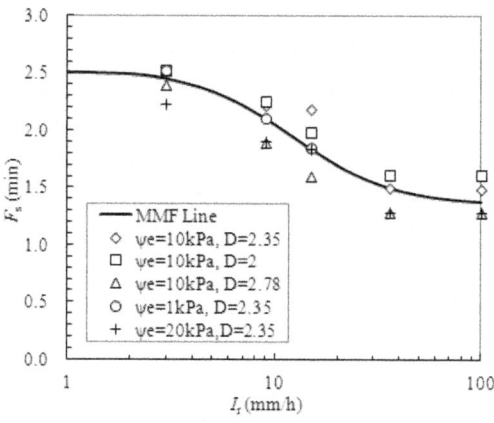

Figure 47. Effect of rainfall intensity on variation of minimum factor of safety.

Effect of initial water table location

The effects of initial groundwater table location on stability of a homogenous soil slope are reflected through the relationship between $F_s(ini)$ and $F_s(min)$ with H_w, shown in Fig. 48. The initial depth of water table, H_w, was varied with five different values of 0.5, 1, 2, 2.5, 5, 7.5, 10 and 15m for a homogeneous soil slope of constant soil type S10, 2.35, -5 ($\psi_e=10kPa$, D=2.35, $k_s=10^{-5}$m/s), $H_s=6m$, $\alpha=33.7°$, subjected to rainfall for 24h with five rainfall intensities of 3, 9, 15, 36, and 100mm/h. The relationship between $F_s(ini)$, and H_w shown in Fig. 48 appears to be linear up to a depth of 7.5m beyond which $F_s(ini)$ remains constant. This is because the initial pore-water pressure profiles generated for the slope at H_w=7.5m as same as those generated when H_w=7.5m. This is due to the limiting pore-water pressure of −75kPa adopted in the analyses. In Fig. 54 it also shows that the rainfall intensity is the dominated factor impacting on the reduction in factor of safety, Fs. And the initial depth of water table, H_w, mainly determines the value of initial factor of safety, F_s(ini). The F_s(ini) is smaller for slopes with a shallower H_w which means that the soil slope stability is much lower. Therefore, slopes with a shallow H_w are more likely to fail due to a rainfall compared with slopes which has a deep H_w.

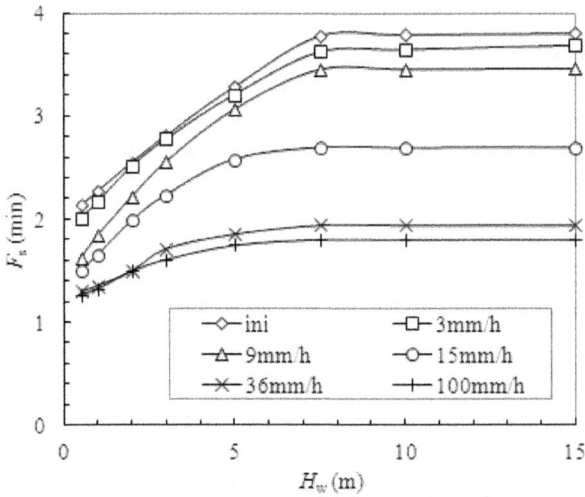

Figure 48. Effect of initial groundwater table depth on factor of safety.

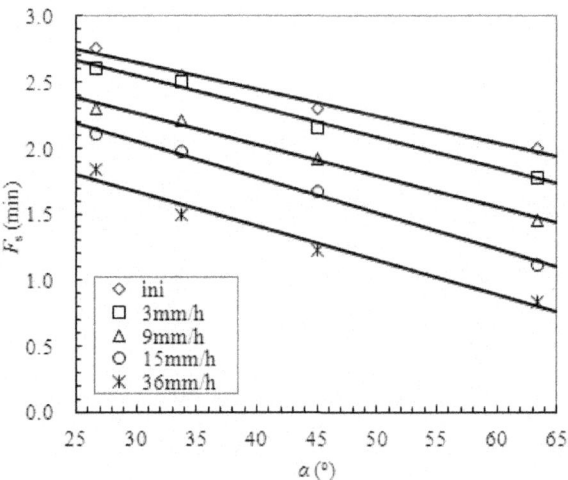

Figure 49. Effect of slope angle on factor of safety.

Effect of Slope Angle

The effect of slope geometry on the stability of a homogenous soil slope is evaluated in terms of slope angle (α) and slope height (H_s). The effects of slope angel on the stability of a homogenous soil slope are reflected through the relationship between $F_s(ini)$ and $F_s(min)$ with α, shown in Fig. 49. The slope angel, α was varied with three different values of 26°, 33.7°, 45°and 63° for a homogeneous soil slope of constant soil type S10, 2.35, -5 (ψ_e=10kPa, D=2.35, k_s=10⁻⁵m/s), H_s=6m, H_w=2m, subjected to rainfall for 24h with five rainfall intensities of 3, 9, 15 and 36mm/h. The relationship between $F_s(min)$ and α shown in Fig. 49 appears to be negative linear. In general, the higher the slope angle, the lower the initial factor of safety and the minimum factor of safety. Because a steep slope will yield a lower factor of safety compared with a flat slope.

Effect of Slope Height

The effects of slope angel on the stability of a homogenous soil slope are reflected through the relationship between $F_s(ini)$ and $F_s(min)$ with H_s, shown in Fig. 50. The slope height, H_s, was varied with four different values of 3m, 4m, 5m and 6m for a homogeneous soil slope of constant soil type S10, 2.35, -5(ψ_e=10kPa, D=2.35, k_s=10⁻⁵m/s), H_w=2m, α=33.7°, subjected to rainfall for 24h with four rainfall intensities of 3mm/h, 9mm/h, 15mm/h and 36mm/h. Fig. 50 shows that initial factor of safety decreases exponentially as the slope height increases. It also suggests that high slopes are generally easier to fail under rainfall due to the low initial factor of safety. The reduction in factor of safety for a high slope is smaller and occurs at a slower rate compared with a low slope.

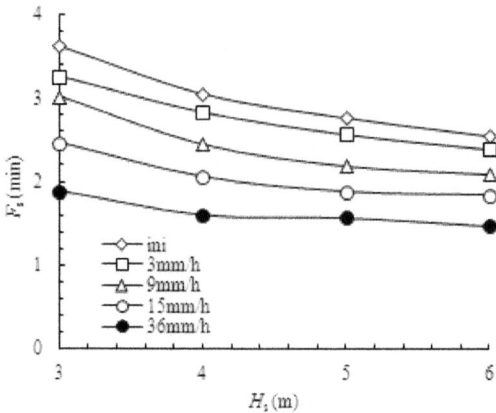

Figure 50. Effect of slope height on factor of safety.

NAN. ACKNOWLEDGEMENTS

The National Natural Science Foundation of China (Grant No. 41272318) and State Key Laboratory of Ocean Engineering are sincerely acknowledged for their financial support.

REFERENCES

1. Abramowitz M, Stegun I. Handbook of Mathematical Functions. Dover Publications, New York, NY. 1970

2. Avnir D, Jaroniec M, An isotherm equation for adsorption on fractal surfaces of heterogeneous porous materials. Langmuir, 1989, 5, 1431–1433.

3. Brooks RH, Corey AT. Properties of porous media affecting fluid flow. ASCE J Irrig Drain Div 1966, 92:61–68.

4. Brutsaert W. Some methods of calculating unsaturated permeability. Trans Am Soc Agr Engrs 1967;10:400-404.

5. Brutsaert W. The permeability of a porous medium determined from certain probability laws for pore size distribution. Water Resour Res 1968; 4: 425-34.

6. Brutsaert W. A concise parameterization of the hydraulic conductivity of unsaturated soils. Advances in Water Resources, 2000, 23, 811-815.

7. Burdine NT. Relative permeability calculations from pore-size distribution data. Trans Am Inst Min Engrs 1953;198:71-7

8. Burdine NT. Relative permeability calculation from pore size distribution data. Transactions of the American Institute of Mining Engineers, 1953, 198, 71–78.

9. Carman PC. Flow of gases through porous media. New York: Academic Press, 1956.

10. Childs EC, Collis-George N. The permeability of porous materials. Proc Roy Soc A 1950; 201: 392-405.

11. Collins,B.D., Znidarcic. Stability analyses of rainfall induced landslides [J].Journal of geotechnical and geoenvironmental engineering, 2004,4,362-372.

12. Corey AT. Pore-size distribution. In: van Genuchten MTh, Leij FJ, Lund LJ, editors. Proceedings of the International Workshop, Indirect Methods for Estimating the Hydraulic Properties of Unsaturated Soils, University of California, Riverside, 1992. p. 37–44.

13. Crawford JW. The relationship between structure and the hydraulic

conductivity of soil. Eur J Soil Sci 1994, 45:493–501.

14. Delage P, Audiguier M, Cui YJ, Howat MD. Microstructure of a compacted silt. Can Geotech J 1996;33:150-158.

15. Dirksen C. Unsaturated hydraulic conductivity. In: Smith KA, Mullins CE, editors. Soil analysis physical methods. New York: Dekker; 1991. p. 209–69.

16. Fatt I, Dykstra H. Relative permeability studies. Transactions of the American Institute of Mining Engineers, 1951, 192: 249-255.

17. Fuentes C, Vauclin M, Parlange J-I. A note on the soil–water conductivity of a fractal soil. Transp Porous Media 1996, 23:31–36.

18. Fuentes C, Haverkamp R, Parlange J-Y. Parameter constraints on closed-form soil water relationships. Journal of Hydrology, 1992, 134, 117-142.

19. Gates JI, Tempelaar-Lietz W. Relative permeabilities of California cores by the capillary pressure method. Drilling and Production Practice, Am Petrol Inst 1950:285-298.

20. GEO-SLOPE International Ltd. Seepage Modeling with SEEP/W2007 [M].Third Edition, Canada, 2008,3.

21. GEO-SLOPE International Ltd. Air Flow Modeling with AIR/W2007 [M].Third Edition, Canada, 2008,3.

22. GEO-SLOPE International Ltd. Stability Modeling with SLOPE/W2007 [M].Third Edition, Canada, 2008,3.

23. Gimenez D, Perfect E, Rawls WJ, Pacheoaky Ya. Fractal models for predicting soil hydraulic properties: a review. Eng Geo 1997;48:161-183.

24. Hoffmann-Riem H, van Genuchten M.Th, Flühler H. A general model of the hydraulic conductivity of unsaturated soils, in: van Genuchten M.Th, Leij FJ, Wu L. (Eds.), Proceedings of International Workshop, Characterization and Measurements of Hydraulic Properties of Unsaturated Porous Media. Riverside, CA. 22–24th Oct. 1997. University of California, Riverside, 1999, pp. 31–42.

25. Hunt AG, Gee GW. Application of critical path analysis to fractal porous media:comparison with examples from the Hanford site. Adv in Water Resources 2002;25:129–46.

26. Irmay S. On the hydraulic conductivity of unsaturated soils. Trans Am Geophys Un 1954;35:463-467.

27. Jarvis NJ, Messing I. Near-saturated hydraulic conductivity in soils of contrasting texture as measured by tension infiltrometers. Soil Sci Soc Am J 1995;59:27–34.

28. Kahr G., Kraehenbuehl F, Stoeckli HF, Muller-Vonmoos M. Study of the

water –bentonite system by vapour adsorption, immersion colorimetry and X-ray technique: Part II. Heats of immersion, swelling pressures and thermodynamic properties. Clay Miner. 1990, 25, 499–506.

29. Khaleel R, Relyea JF. Evaluation of van Genuchten–Mualem relationships to estimate unsaturated hydraulic conductivity at low water contents. Water Resour Res 1995; 31: 2659–1668.

30. Kozeny J. Ueber kapillare leitung des wassers im boden, sitzungsberichte, akad der wissensch. Wien, Math-Naturw Klass Abt IIa 1927;136:271-306.

31. Mandelbrot BB, The Fractal Geometry of Nature, W.H. Freeman, New York, 1982.

32. Matsuoka H, Soil Mechanics. Morikita Shuppan Co., Ltd., 1999, p. 61.

33. Mualem Y. A new model for predicting the hydraulic conductivity of unsaturated porous media. Water Resources Res 1976, 12:513–22.

34. Polubarinova-Kochina PYa. Theory of ground water movement. Princeton, NJ: Princeton University Press, 1952. p. 613 (translated from the Russian by DeWiest, JMR 1962).

35. Purcell WR. Capillary pressures-their measurement using mercury and the calculation of permeability therefrom. Trans Am Inst Min Met Engrs Petrol Devel Technol 1949;186:39-46.

36. Rahardjo H, Ong TH, Rezaur RB. Factors controlling instability of homogeneous soil slopes under rainfall [J]. Journal of geotechnical and geoenvironmental engineering, 2007,12,1532-1543.

37. Ravichandran N, Krishnapillai S. A Statistical Model for the Relative Hydraulic Conductivity of Water Phase in Unsaturated Soils. International Journal of Geosciences, 2011, 2, 484-492.

38. Rieu M, Sposito G. Fractal fragmentation, soil porosity and soil water properties. Soil Sci Soc Am J 1991; 55: 1483–1489.

39. Schaap MG, Leij FJ, van Genuchten MT. Rosetta: A computer program for estimating soil hydraulic parameters with hierarchical pedotransfer functions, J. Hydrol., 2001, 251, 163-176.

40. Simunek T, Wendroth O, van Genuchten MT, et al., Water Resources Res. 35 (1999) 2965.

41. Smettem KRJ, Kirkby C. Measuring the hydraulic properties of a stable aggregate soil. J Hydro 1990, 117: 1-13.

42. Stingaciu LR, Weihermüller L, Haber-Pohlmeier S et al. Determination of pore size distribution and hydraulic properties using nuclear magnetic resonance relaxometry: A comparative study of laboratory methods.

Water Resources Research, 2009, 46: W11510.

43. Sugii T, Uno T, Hayashi T, Proceedings of the 31st Japan National Conference on Geotechnical Engineering, Kitami, Japan, 1996, pp. 2075

44. Toledo PG, Novy RA, Davis HT, Scriven LE. Hydraulic conductivity of porous media at low water content. Soil Sci Soc Am J 1990;54:673–9.

45. Touma J, Vauclin M. Experimental and numerical analysis of two-phase infiltration in a partially saturated soil. Transport in Porous Media, 1986, 1, 27-55.

46. Touma J. Comparison of the soil hydraulic conductivity predicted from its water retention expressed by the equation of van Genuchten and different capillary models. European Journal of Soil Science, 2009, 60: 671-680

47. Tyler SW, Wheatcraft SW. Fractal process in soil water retention. Water Res Res 1990, 26:1047–54.

48. Uno T, Kamiya K, Tanaka K. The distribution of sand void diameter by air intrusion method and moisture characteristic curve method. Proc Japan Soc Civil Engrg, 1998, 603(III-44): 35-44.

49. van Genuchten MTh, Leij FJ. On estimating the hydraulic properties of unsaturated soils. In: van Genuchten M.Th, Leij FJ, Lund LJ, editors. Proceedings of the International Workshop, Indirect Methods for Estimating the Hydraulic Properties of Unsaturated Soils, University of California, Riverside, 1992. p.1-14.

50. van Damme H. Scale invariance and hydric behaviour of soils and clays. CR Acad Sci Paris 1995, 320: 665–81.

51. van Genuchten MTh. A close form equation for predicting the hydraulic conductivity of unsaturated soils. Soil Sci Soc Am J 1980;44:892–7.

52. Vogel HJ, Roth K. A new approach for determination effective soil hydraulic function. Eur J Soil Sci 1998;49:547–56.

53. Watabe Y, Leroueil S, Le Bihan J-P, Influence of compacted conditions on pore-size distribution and saturated hydraulic conductivity of a glacial till, Can Geotech J, 2000, 37:1184-1192.

54. Wheeler SJ, Karbue D, Constitutive modeling, In: Alonso EE, Delage P, Proc. 1st Int Conf Unsat Soils Rotterdam: AA Balkema, 1996.

55. Wyllie MRJ, Spangler MB. Application of electrical resistivity measurements to problem of fluid flow in porous media. Bull Am Ass Petrol Geologists 1952, 36:359-403.

56. Xu YF, Sun DA. A fractal model for soil pores and its application to determination of water permeability. Physica A 2002;316(1-4): 56–64.

57. Xu YF, Dong P. Fractal approach to hydraulic properties in unsaturated

porous media. Chaos, Solitons & Fractals, 2004, 19(2): 327-337

58. Xu YF, Sun DA, Yao YP. Surface fractal dimension of bentonite and its application to determination of swelling properties. Chaos, Solitons & Fractals, 2004, 19(2): 347-356.

59. Xu YF, Calculation of unsaturated hydraulic conductivity using a fractal model for the pore-size distribution, Computer and Geotechnics, 2004, 31(7):549-557.

60. Xu YF. Fractal approach to unsaturated shear strength. Journal of Geotechnical and Geoenvironmental Engineering, 2004, 130(3): 264-273

Chapter 2

THE IMPACT OF ICE COVER AND SEDIMENT NONUNIFORMITY ON EROSION AROUND HYDRAULIC STRUCTURES

Peng Wu[1], Jueyi Sui[2] and Ram Balachandar[3]

[1] Environmental Systems Engineering, University of Regina, Canada

[2] Environmental Engineering Program, University of Northern British Columbia, Canada

[3] Civil and Environmental Engineering, University of Windsor, Canada

ABSTRACT

Based on two case studies, the impact of ice cover on local scour around bridge piers is presented in this chapter. Bed material with different grain sizes is used and ice covers with different roughness is used to study the scour characteristics. The impact of nonuniformity of sediment is also investigated. Results show that with the increase in densimetric Froude number, there is a corresponding increase in the dimensionless scour depth. For nonuniform sediment, due to the formation of an armor layer, less maximum scour depth was noted around bridge foundation structures compared to uniformly distributed sediment. The increase in ice cover roughness results in a larger scour depth and geometry. The results indicate that it is imperative to pay attention to the impact of ice cover on the scour around hydraulic structures.

INTRODUCTION

SEDIMENT TRANSPORT AROUND HYDRAULIC STRUCTURES

Sediment transport embodies the processes of erosion, entrainment, movement, and deposition. In nature, these processes are always present. Although there are many types of sediment, fluvial sediment is one of the most common. It comprises sediment accumulated by the erosion of rock and mineral particles that are transported by flowing water. The process of erosion is complex and plays a vital role in the formation of different landscapes of the world we live in.

Statistics show that thirteen of the large rivers in the world carry sediment loads in excess of 5.8 billion tons annually [5]. In addition to producing large quantities of sediment, erosion also causes other problems such as the pollution of water body and altering the runoff conditions. Additionally, from the perspective of hydraulic engineering, sediment transport can result in serious on-site damage to hydraulic structures such as bridge abutments, bridge piers, spur dikes, etc.

The presence of a hydraulic structure in a river introduces the phenomenon of local scour. The interaction between flow and structures has been an actively researched topic in the past decade. Reynolds stress is an important parameter to quantify the suspended load of sediment transport, while the bed shear stress is pertinent to determine the bed load of sediment transport. Kuhnle et al. [14] examined the three-dimensional flow field around a submerged spur dike. An amplification factor of bed shear stress 3 was found. Duan et al. [7] measured the flow and turbulence around an experimental spur dike in a flat and scour bed. Differences of mean velocity, turbulent intensity, and Reynolds stresses between those two flow fields were analyzed. The bed shear stress calculated from the Reynolds stresses was 2-3 times that of the incoming flow. Since the abutment and spur dike have similar contraction impact on the flow, all the three studies showed similar amplification factor of bed shear stress in open channel flow.

Impact of ICE Cover

Extensive studies have been conducted to study the local scour around structures in the past few decades. However, most of the previous studies were focused on flow in open channels. The impact of ice cover on local scour has not been well understood. For many regions in the northern hemisphere, winter may last up to six months. Figure 1 shows the ice cover around Ambassador Bridge, Windsor, Canada, which is located on the Detroit River. It can be seen from Figure 1 that the region around the pier is completely covered by ice. The presence of the ice cover affects the flow characteristics of the river. Hence, further research on the impact of ice cover on the local scour around bridge piers becomes important.

As pointed out by Morse and Hicks [21], fundamental problems of ice engineering are primarily related to ice jams, floods, and transportation over ice. River ice hydrology has been an implicit part in ice engineering. Some studies on ice engineering research can be found in the references

[1,12,22,9,21,23,26,30,32]. To date, the mechanism of local scour around bridge abutments and piers under ice cover is still not well understood.

Lau and Krishnappan [15] used the $k-$ turbulence model to numerically determine the velocity distribution and suspended load transport under ice cover. The assumption was that the ice covered flow can be treated as two-layer flow which is divided at the location of maximum velocity. Based on the assumption that the mechanics of bed-form formation is the same for open and ice covered channels, Smith and Ettema [27] developed a method from flume data for estimating the flow resistance in ice covered channels. Ettema et al. [8] proposed a different approach to estimate the sediment transport under ice cover by using the procedures from open channel flow. The variation of ice cover roughness and flow cross sections under ice cover can limit the accuracy of this method. Ettema and Daly [9] conducted a series of flume experiments on the sediment transport under ice cover. Since the flow distribution is substantially modified by ice cover, the sediment transport under ice cover is hard to estimate. Sui et al. [26] conducted a series of flume experiments for the incipient motion under different flow and boundary conditions. The influence of ice cover roughness has been assessed. They found that the slope of ice cover has a great impact on the critical dimensionless shear stress for the river sediment motion.

Figure 1. The ice cover around bridge piers (Ambassador Bridge, Windsor, ON, Canada, 2015).

Impact of Nonuniformity of Sediment

Incipient motion of the particles is an important criterion that determines the motion of the sediment particles. When the flow attains or exceeds the criteria for incipient motion, sediment particles along the alluvial channel start to move. As defined by Yang [33], if the motion is rolling, sliding, or jumping along the bed, it is called bed load transport. If the particle is supported by the upward components of turbulent currents and stays in suspension, it is called suspended load transport.

For nonuniform sediment, no single critical grain diameter can be determined to distinguish which size of sediment moves and which does not for any specific conditions. Therefore, the incipient motion of bed material is also the beginning of armor process of the bed surface [5]. With the development of an armor layer, further sediment transport is inhibited. Since nonuniform sediment makes up the typical bed composition in natural rivers, the study of the impact of nonuniformity has to be considered with the formation of armor layer.

The forces acting on a sediment particle at the bottom of the scour hole under ice cover are shown in Figure 2. In natural rivers, the velocity profile is similar to the one shown in Figure 2, which exaggerates the impact of ice cover. Herein, the figure is used to show the impact of ice cover on velocity profile around bridge piers and abutments.

For most natural rivers, the river slopes are small enough that the component of gravitational force acting on the particle in the direction of flow can be neglected. As shown in Figure 2, the forces to be considered related to the incipient motion are the drag force FD, lift force FL, submerged weight W, and the resistance force FR. The angle of the scour hole with vertical abutment is .

A sediment particle is at a state of incipient motion when the following conditions have been satisfied:

$$\begin{cases} F_D = F_R \sin \alpha \\ W = F_L + F_R \cos \alpha \end{cases}$$

$$(1)$$

From existing literature it is apparent that studies on local scour around hydraulic structures under ice covered conditions with nonuniform sediments are limited. The effects of ice cover and nonuniformity have to be considered in the analysis of local scour. In this chapter, Equation (1) and Figure 2 are used as the basis for analysis of incipient motion under ice cover. The force analysis of a particle under ice cover is conducted by introducing the armor layer particle size. At the end, dimensionless shear stress is calculated by using ADV (Acoustic Doppler Velocimetry) data.

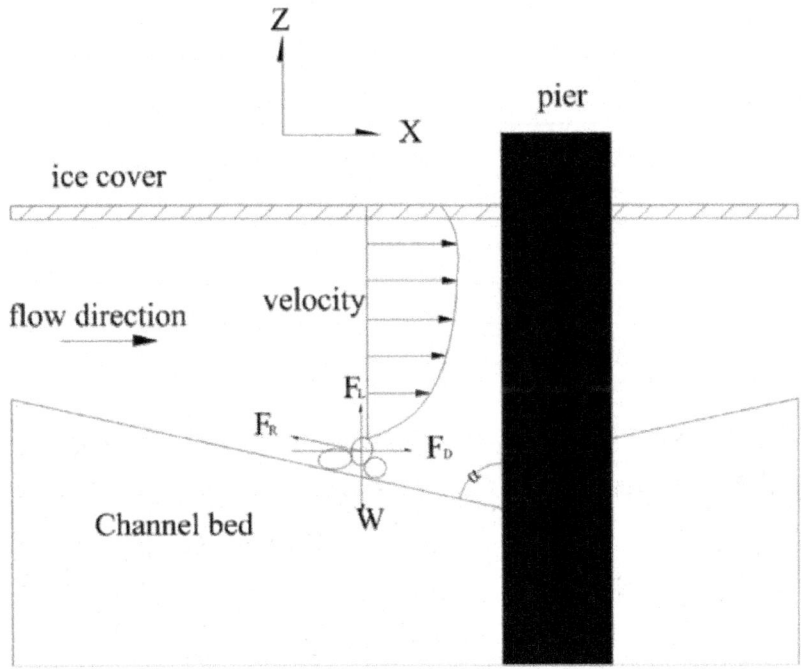

Figure 2. Incipient motion in the scour hole under ice cover (modified by [30]).

CASE STUDIES

Experimental Setup and Measurement

The first case study was conducted in a 40-m long, 2-m wide, and 1.3-m deep flume with two sand boxes. The flume is a large-scale flume located at Quesnel River Research Center, BC, Canada. The aim of this case study was to investigate the ice cover impact on bridge abutment with nonuniform sediment. Smooth and rough ice covers were used. Styrofoam is commonly used to simulate ice cover in both case studies. Two different nonuniform sediments were used with D_{50} of 0.58 mm and 0.47 mm, respectively. The geometric standard deviations (σ_g) were all larger than 1.4 for the sediments, hence they can be treated as nonuniform sediment. The D_{90} of the two sediments was 2.57 mm and 1.19 mm, respectively. The sieve analysis of sediments can be found in Figure 3. Square and round abutment models were made with an equivalent diameter of 200 mm. A constant blockage of 10% was used for all experiments. Table 1 summarizes the running conditions. In all, 36 experiments were conducted.

Table 1. Summary of Case 1.

Abutment type	Cover condition	D_{50}(mm)	Water depth (m)	Approaching velocity (m/s)
Square	Open	0.58	0.07	0.26
Square	Open	0.58	0.07	0.21
Square	Open	0.58	0.19	0.21
Round	Open	0.58	0.07	0.21
Round	Open	0.58	0.19	0.23
Round	Open	0.58	0.07	0.26
Round	Smooth	0.58	0.07	0.23
Round	Smooth	0.58	0.19	0.20
Round	Smooth	0.58	0.07	0.20
Square	Smooth	0.58	0.07	0.20
Square	Smooth	0.58	0.19	0.16
Square	Smooth	0.58	0.07	0.23
Square	Rough	0.58	0.07	0.22
Square	Rough	0.58	0.07	0.20
Square	Rough	0.58	0.19	0.14
Round	Rough	0.58	0.07	0.20
Round	Rough	0.58	0.19	0.20
Round	Rough	0.58	0.07	0.22
Square	Open	0.47	0.07	0.26
Square	Open	0.47	0.07	0.21
Square	Open	0.47	0.19	0.21
Round	Open	0.47	0.07	0.21
Round	Open	0.47	0.19	0.23
Round	Open	0.47	0.07	0.26
Round	Smooth	0.47	0.07	0.23
Round	Smooth	0.47	0.19	0.20
Round	Smooth	0.47	0.07	0.20
Square	Smooth	0.47	0.07	0.20
Square	Smooth	0.47	0.19	0.16
Square	Smooth	0.47	0.07	0.23
Square	Rough	0.47	0.07	0.2
Square	Rough	0.47	0.07	0.20

Square	Rough	0.47	0.19	0.14
Round	Rough	0.47	0.07	0.20
Round	Rough	0.47	0.19	0.20
Round	Rough	0.47	0.07	0.22

Table 2. Summary of Case 2 (H represents flow depth; D is the pier diameter).

H (mm)	Surface condition	Test Number	D (mm)	H/D	$D/_{DS}0$
108	Open channel	A1	16	6.8	31
		A2	30	3.6	59
		A3	42	2.6	82
		A4	90	1.2	176
	Ice cover	B1	16	6.8	31
		B2	30	3.6	59
		B3	42	2.6	82
		B4	90	1.2	176

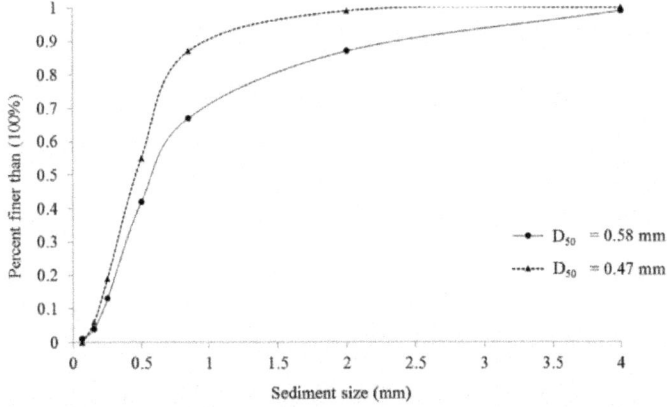

Figure 3. Sieve analysis of sediments used in Case 1.

In each experiment, a 10 MHz SonTek ADV was used to measure the 3D flow velocity in the vicinity of abutments. For the ADV measurement, two values were used to ensure the measurements can provide an accurate representation of the flow velocity: signal-to-noise ratio (SNR) larger than 15 db and the correlation (COR) between 70% and 100%. Then, the data were analyzed by WinADV (Wahl, 2000). However, due to the limitations of ADV, the velocity profile close to the ice cover could not be measured.

The second case study was conducted at the Sedimentation and Scour Study Laboratory at the University of Windsor. The flume has a dimension of 12 m long, 1.2 m wide. The aim of this study was to investigate the ice cover impact on scour around a bridge pier. Only one type of ice cover and one uniform sediment with D_{50} of 0.51 mm was used. The sieve analysis of sediment is presented in Figure 4. Four pier diameters and one water depth was selected for comparison. In all, 8 experiments were conducted. Table 2 summarizes the test conditions in the case study. This study aims to investigate the ice cover impact around bridge piers with uniform sediment. A test period of 48 hours was selected to achieve near-equilibrium conditions. After each experiment, the scour profile and scour pattern were measured by using a laser point gauge.

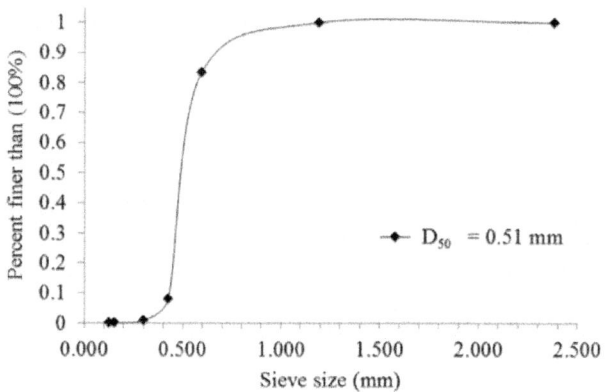

Figure 4. Sieve analysis of sediments used in Case 2.

Results and Analysis

Case 1

Figure 5 shows the scour pattern around abutments with nonuniform sediment. As shown in the figure, armor layer can be clearly noted on top of the scouring region around abutments. In both upstream and downstream of the abutment, coarse sediments were distributed equally around the scouring area. From experimental observation, the formation of armor layer greatly affects the maximum scour depth. Due to armor layer, the time needed to reach maximum scour depth is less than 24 hours.

Typical scour patterns around square and round abutments under different covered conditions were mapped. Figure 6 shows the contours around round abutment under both smooth cover and rough cover in the same flow conditions.

Under both the covered conditions, the maximum scour depth is located at 60–70 degrees facing upstream. Additionally, rough cover can yield a relative larger scour compared to that from smooth cover as shown in Figure 6.

In all, 36 maximum scour depths are plotted in Figure 7, in which 18 are from a square abutment and 18 from a round abutment. Figure 7 shows the difference in maximum scour depth between the two types of abutments under different flow conditions. Based on Melville [19], the shape factor for square abutment in open channels is assumed as 1.0, while for the abutment with a round head, the value is assumed as 0.75. Figure 7 indicates that under both open channel and ice-covered conditions, square abutment generates a relatively larger maximum scour depth compared to the round abutment.

In order to compare the difference between square and round abutment, densimetric Froude number (F_o) can be calculated by using the following:

$$F_o = U_o / \sqrt{g(\Delta \rho / \rho) D_{50}}$$

(2)

where g is the gravitational acceleration, U_o is the approaching velocity, ρ is the mass density of water, while the $\Delta \rho$ is mass difference between sediment and water. D_{50} is the median grain size of sediments. An analysis using the densimetric Froude number is conducted to quantify the impact of abutment shape on maximum scour depth.

Furthermore, ice cover has a significant impact on the dimensions of the scour hole. In open channels, around the square abutment, the scour hole has a smaller slope compared to the scour holes under ice cover. It is also interesting to note that around the round abutment, the area of scour hole under the ice cover is larger than that of open channel. With the increase in ice cover roughness, there is a corresponding increase in scouring area. Around the square abutment, the scour hole retains a similar pattern with or without ice cover. While for round abutment, the scour hole under ice cover is larger than that from smooth cover and open channel.

To further examine the role of the ice cover, Figure 8 is plotted.

From Figure 8, at least two observations can be noted. Firstly, under all cover conditions, the overall trend of maximum scour depth increases with the increase in densimetric Froude number. Secondly, for a given F_o, the rough ice cover yields the largest maximum scour depth, while open channel has the smallest value. The data in Figure 8 also indicates the impact of ice cover on maximum scour depth. As mentioned previously, armor layer was detected in the scouring area around abutments.

Figure 5. Scour pattern with nonuniform sand from Case 1.

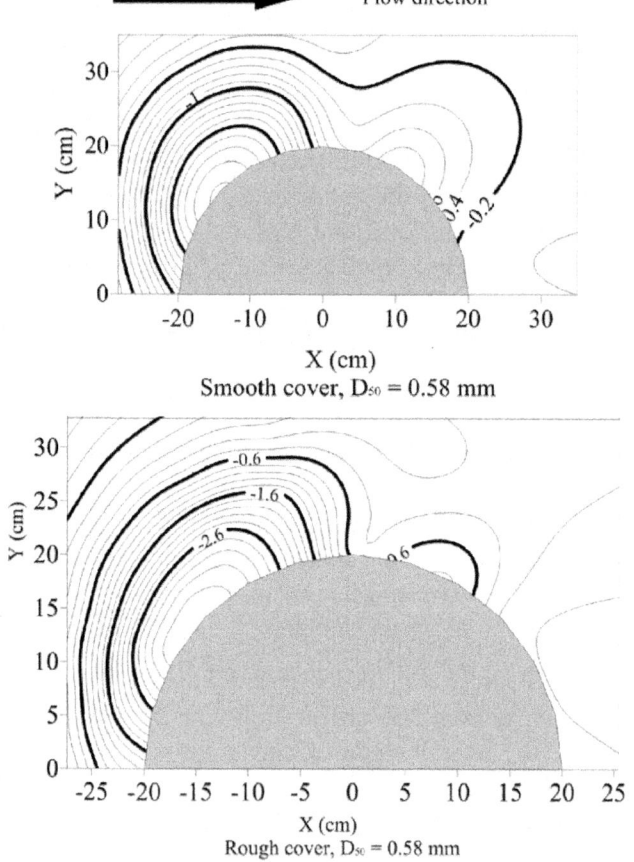

Figure 6. Scour profile around round abutment.

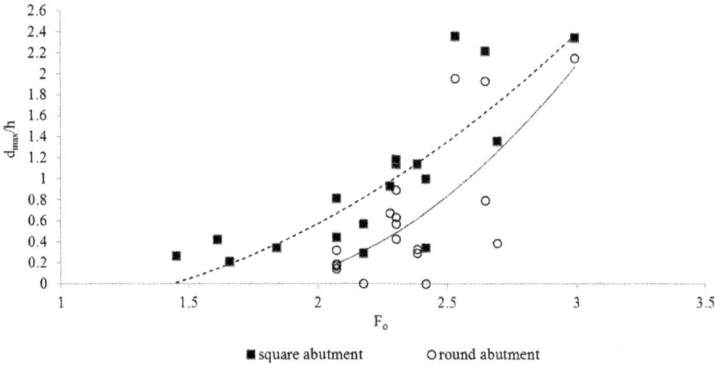

Figure 7. The dimensionless maximum scour depth variation with densimetric Froude number.

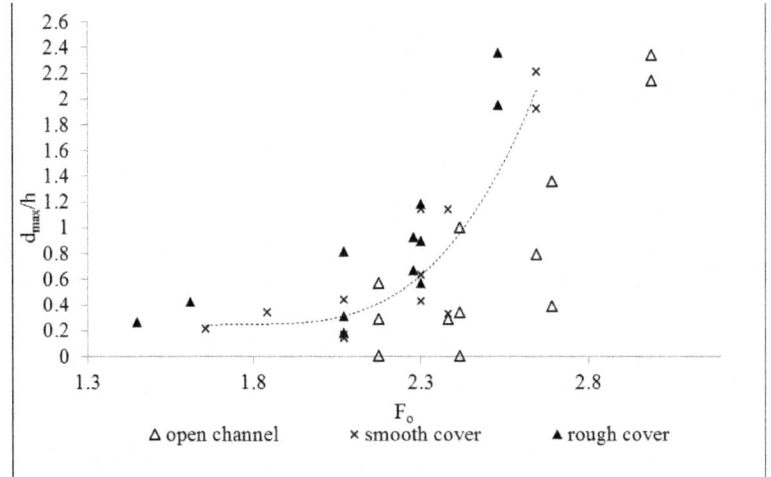

Figure 8. The comparison of dimensionless scour depth under different cover conditions (h is the flow depth).

Figure 9 shows armor layer sieve analysis compared with the original sediment sieve analysis. From the photos, it is interesting to note the proportion of coarse particles in the scouring area. Additionally, comparing the two analysis curves, it should become apparent that a smaller D_{50} yields small difference between armor layer and original sand, because smaller D_{50} for nonuniform sediment implies less coarse particles, hence the sediment size in the armor layer would decrease accordingly. Less coarse particles in the armor layer will provide less protection to foundation. Hence, a relatively larger maximum scour depth would be formed.

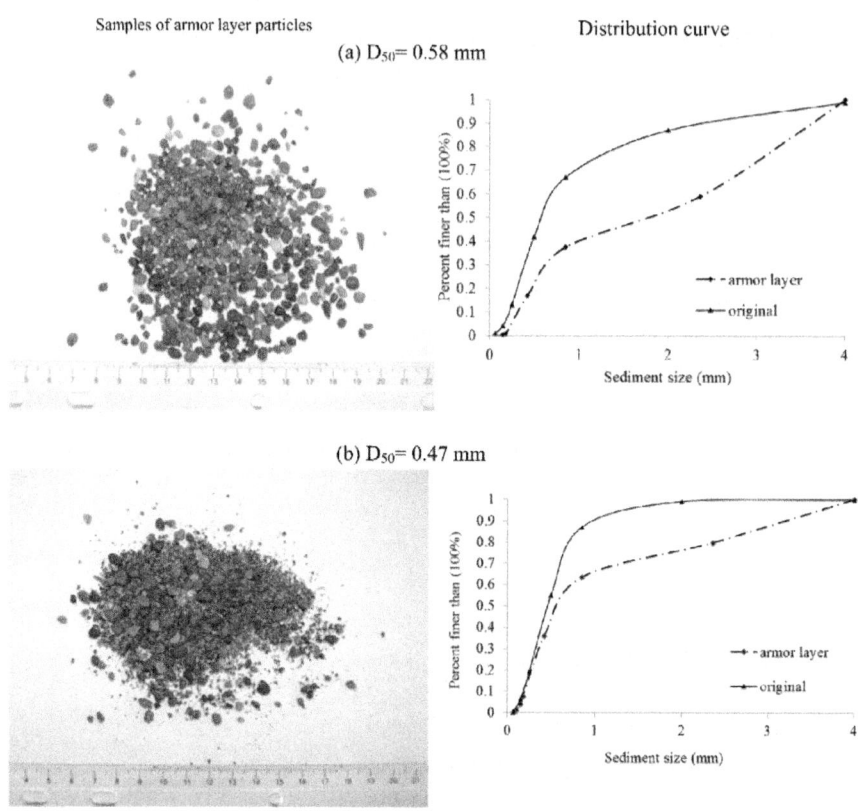

Figure 9. Samples of armor layer particle size and distribution curves.

Dimensional analysis provides a convenient way for building a framework for parameters on which the maximum scour depth depends. Given the complexity of the interaction of various parameters, NCHRP [24] identified five major groups of dimensionless parameters affecting the maximum scour depth. However, there is still no clear indication on the impact of nonuniformity. In other words, the ice cover impact was not considered. Therefore, it is necessary to conduct a dimensionless analysis to investigate the ice cover impact on maximum scour depth. The following relationship can be noticed regarding maximum scour depth:

$$d_{\max} = f(U, \rho, \rho_s, g, d, n_b, n_i, D_{50}, l, B, H)$$

(3)

where

$d_{\max}$ is the maximum scour depth around the abutment,

U is the mean approach velocity,

ρ is the mass density of the water,

ρ_s is the mass density of nonuniform sediment,

d is the armor layer grain size,

n_b is the Manning's coefficient for the channel bed,

n_i is the Manning's coefficient of ice cover,

D_{50} is the median grain size,

l is the abutment width,

B is the flume width, and

H is the approaching water depth.

In the above relationship, the terms g, ρ, and ρ_s can be eliminated by introducing a combining parameter for a flow-sediment mixture. Additionally, abutment blockage ratio is also kept constant in the case study. To include the impacts of nonuniformity and armor layer, Meyer-Peter and Müller [20] developed the following equation by using one mean grain size of the bed to calculate the sediment size in the armor layer:

$$d = \frac{SD}{K_1 \left(n/D_{90}^{1/6} \right)^{3/2}}$$

$$(4)$$

where

d is the armor layer sediment size,

S is the channel slope,

D is the mean water depth,

K_1 is the constant number, which equals to 0.058 when D is in meter,

n is the channel bottom Manning's roughness, and

D_{90} is the bed material size where 90% of the composition is finer.

Since the armor layer particle size is of the main interest here, d is introduced in the calculation of the densimetric Froude number:

$$F_o' = U/\sqrt{(\rho_s/\rho - 1) gd}$$

$$(5)$$

Figure 10 shows the variation of dimensionless maximum scour depth d_{max}/d with F_o' around both square and round abutments under all three flow conditions. It should be noted that under all flow conditions, with the increase in F_o', the value of d_{max}/d increases correspondingly. Around both square and round abutments, under the same F_o', the rough ice cover has the largest dimensionless maximum scour depth. Smooth ice cover has the second largest value. Meanwhile, the largest dimensionless maximum scour depth is located around square abutment in all flow conditions.

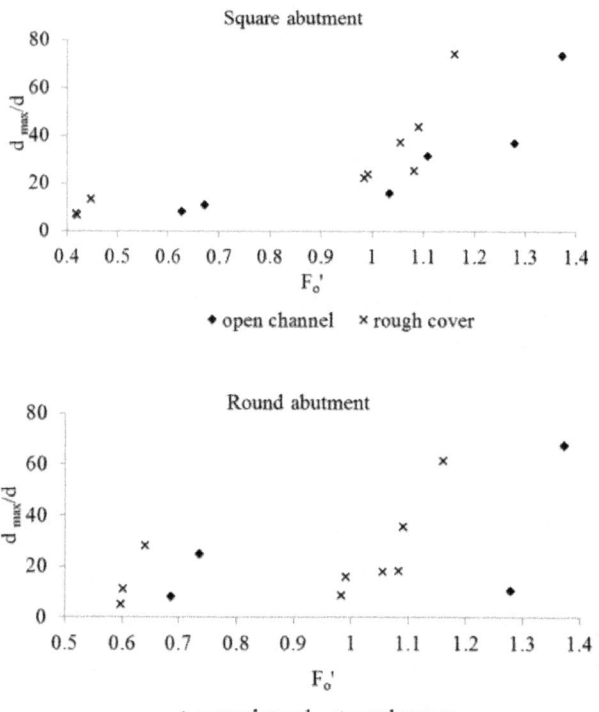

Figure 10. The dimensionless Froude number around both abutments under different cover conditions.

From the above analysis, it is clear that ice cover has strong influence on local maximum scour depth around abutments. But rough ice cover has an even stronger impact.

Manning's roughness coefficient is between 0.01 and 0.0281 based on supporting field data as well as the observed ice cover characteristics [4]. In the case study, the Manning's coefficient of 0.013 was adapted in accordance with Mays [18] for smooth ice cover. Rough ice cover was created by attaching small cubes with dimensions of 2.5 cm × 2.5 cm × 2.5 cm. The following equation was applied to calculate the Manning's coefficient [16]:

$$n_i = 0.039 k_s^{1/6}$$

(6)

where

ks is the average roughness height of the ice underside.

In the first case study, rough ice cover has a Manning's coefficient of 0.021, which is located in the range as suggested by Carey [4]. The channel bed roughness is calculated by using the following equation from Hager [11]:

$$n_b = 0.039 D_{50}^{1/6}$$

(7)

As suggested by Wu et al. [30], the following relationships were developed to show the maximum scour depth under ice cover around square and round abutments:

$$\left(\frac{d_{max}}{d}\right)_{square} \sim (F_o)^{3.73} \left(\frac{D_{50}}{d}\right)^{-1.78} \left(\frac{n_i}{n_b}\right)^{0.77} \left(\frac{H}{d}\right)^{3.01}$$

(8)

$$\left(\frac{d_{max}}{d}\right)_{round} \sim (F_o)^{8.60} \left(\frac{D_{50}}{d}\right)^{-4.30} \left(\frac{n_i}{n_b}\right)^{1.00} \left(\frac{H}{d}\right)^{3.00}$$

(9)

The exponent of each parameter from the above relationships can be used to indicate the potential impact. The exponent of F_o, n_i/n_b, H/d is positive, while the exponent of D_{50}/d is negative. From the previous discussion, densimetric Froude number has a positive impact on maximum scour depth. Moreover, compared to the ice cover roughness, armor layer particle size also has a strong impact on the dimensionless maximum scour depth. With the increase in armor layer particle, the maximum scour depth decreases. This conclusion is in line with a previous study conducted by Sui et al. [26]. In natural river engineering, a mixture of coarse sediments in the vicinity of bridge foundations has the potential to reduce maximum scour depth.

Case 2

Figure 11 shows the impact of ice cover on bridge pier. Only one type of ice cover was used. Some of the results are presented by Wu and Balachandar [29]. At certain water depth, in this case, H = 108 mm, the scour contour and scour profile in the flow direction under ice cover is always larger than that in an open channel. The scour profiles in the flow direction also share some similarity. For example, the scour profiles almost overlap with each other in the upstream direction. While in the downstream of pier, the ice cover can clearly result in a movement of ridge toward downstream.

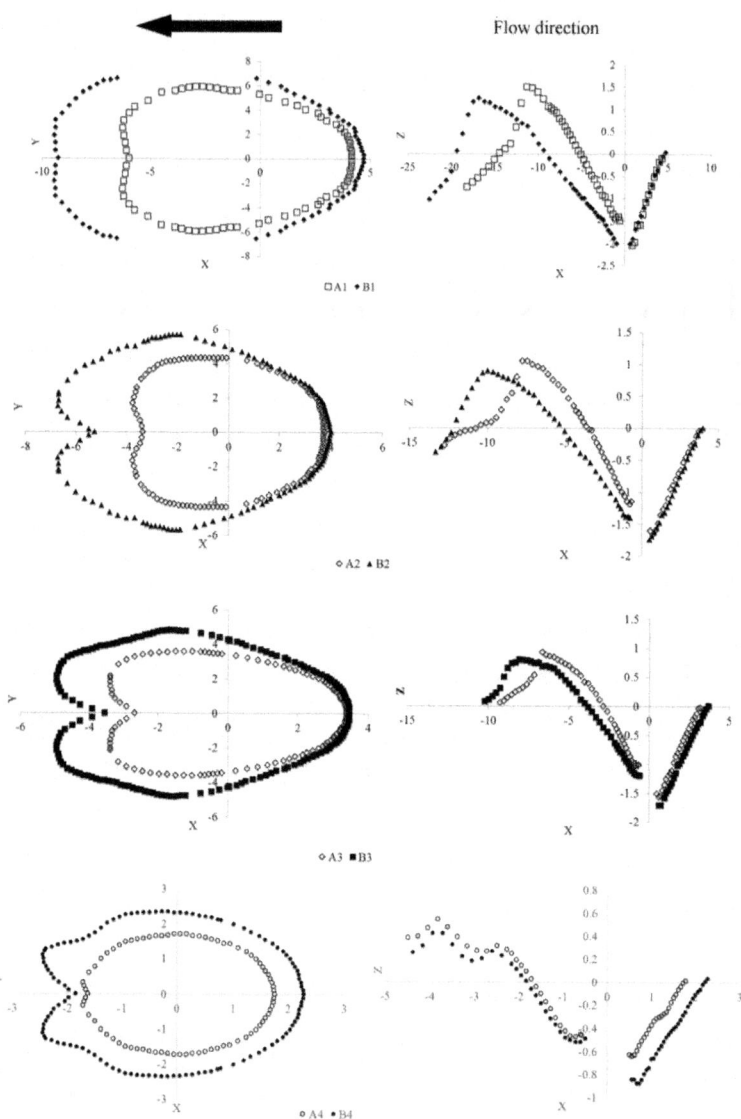

Figure 11. Scour comparison around piers under open channel (Group A) and ice cover (Group B) [29].

From the graphics presented above, the impact of ice cover on the local scour at bridge pier can be studied. It is evident from both case studies that, with the presence of ice cover, the maximum scour depth as well as the scour profile are increased.

Velocity Profile under Ice Cover

By using the ADV data, the instantaneous velocity around the square abutment inside the scour hole was analyzed. The maximum scour depth is located at the upstream corner for the square abutment. After each experiment, the velocity profile was measured at the upstream corner. Here, the parameter of interest is the maximum scour depth, so the velocity profile at the maximum scour depth location was plotted. The three-dimension velocity u, v, w in open channel, under both smooth ice cover, and rough ice cover are plotted in Figures 12 and 13.

For different sediments, the velocity profiles in the scour hole are slightly different. From these figures, one can notice the following:

- In open channel, the velocity in the X direction (flow direction) has the largest magnitude, while the velocity in Y direction (transverse direction) is second largest. One can also observe in open channel, the velocity in the X and Y directions changes direction from positive to negative or from negative to positive. Compared to the change of velocity direction in X and Y directions under ice cover, the open channel has the most turbulence in both the X and Y directions.

- Under ice-covered condition, the velocity magnitude in the Y direction has the largest magnitude followed by the velocity magnitude in X direction. This trend is clear in Figure 13 under both smooth and rough ice cover. At the same flow depth, the velocity in the Y direction inside the scour hole under rough ice cover has the largest value.

- Due to the impact of ice cover on the flow, the velocity in the Z direction is significantly different compared to that of open channels. Under ice-covered condition, the magnitudes in Z direction are larger than that from open channels. While under rough ice cover, the velocity gradient in Z direction is the largest. With decreasing D_{50}, the velocity in Z direction increases correspondingly. This is attributed to the presence of ice cover, which pushes the flow down toward the channel bed. A larger scour depth is formed. With the increase in ice cover roughness, the velocity in Z direction is increased correspondingly. With the decrease in D_{50} and increase in ice cover roughness, the variation of velocity in Z direction is obvious.

- Above the scour hole, the velocity gradients in both X and Y directions are decreased. The maximum velocity in X and Y direction is located at about the mid-depth from bottom of scour hole to the ice cover as shown in Figure 13.

Shear Stress Analysis

Bed shear stress and shear velocity are fundamental variables in river hydraulics to calculate the sediment transport, scour, and deposition. Dey and Barbhuiya [6] developed the Reynolds stresses method which is widely used by engineers. Biron et al. [3] compared several methods for bed shear stress calculation. For complex flow calculation, turbulent kinetic energy (TKE) method provided the best estimate of bed shear stress as it is not affected by local streamline variation and it takes into account the increased turbulent fluctuations. Using ADV data, Biron et al. [3], MacVirar and Roy [17], and Acharya [2] concluded that the TKE method is the most reliable to estimate the bed shear stress. Acharya [2] employed the TKE method for calculating bed shear stresses around spur dikes, yielding satisfactory results.

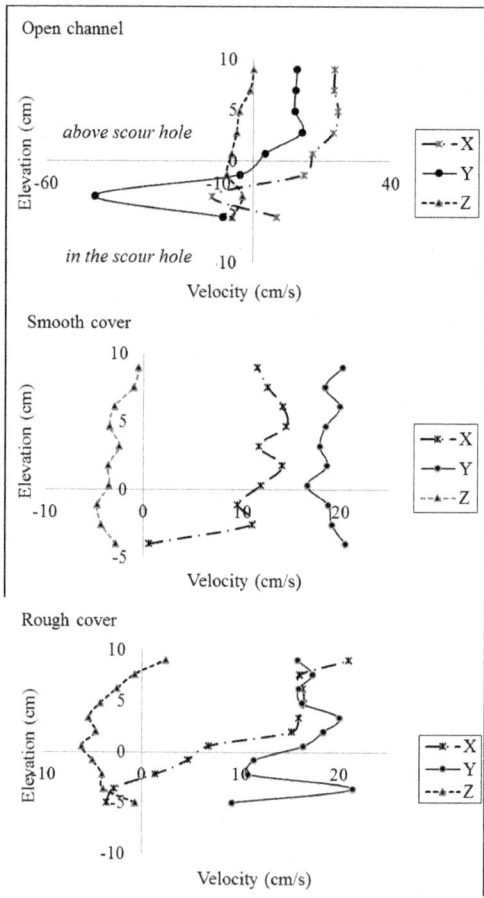

Figure 12. Three-dimensional velocity profile under different cover conditions (*D50* = 0.58 mm).

Near-bed velocities were measured by placing the ADV probe in the scour hole. Three-dimensional velocity vectors were collected. For the instantaneous velocity component u, v, w in the X, Y, Z direction, the time-averaged velocity can be given as:

$$u = \bar{u} + u', v = \bar{v} + v', w = \overline{w} + w'$$
(10)

where u′, v′, w′ denote the turbulent fluctuations in X, Y, and Z directions. The collected velocities at each point were processed to calculate the mean flow and turbulence characteristics at each point.

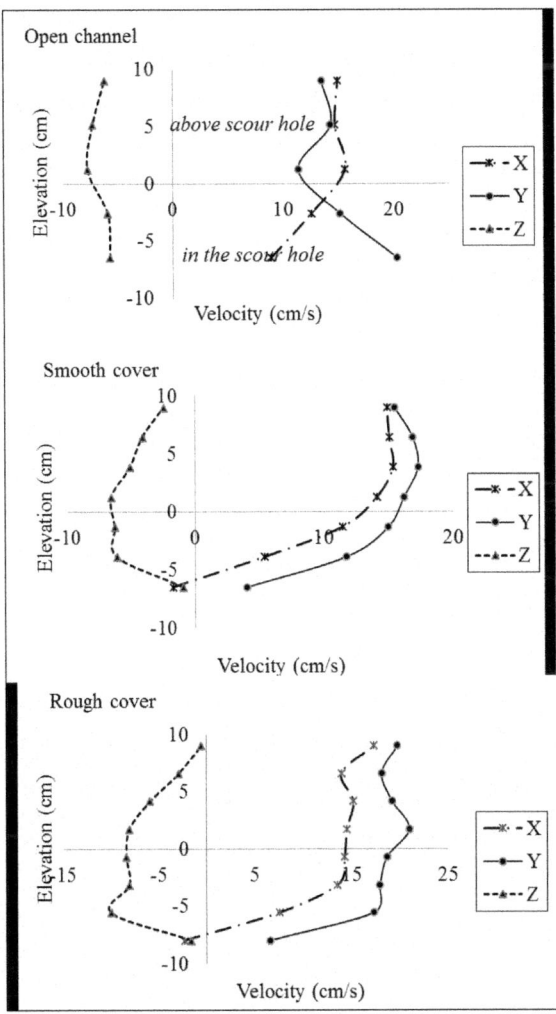

Figure 13. Three-dimensional velocity profile under different cover conditions (D_{50} = 0.47 mm).

The bed shear stress was calculated by using TKE method as:

$$\tau_b = C_1 \left[\frac{1}{2}\rho \left(u'^2 + v'^2 + w'^2 \right) \right]$$

(11)

where C_1 is a proportionality constant which equals 0.19 [13]. Due to the fact that instrument noise errors associated with vertical velocity fluctuations were smaller than that of the horizontal velocity fluctuations, the above equation then can be simplified as:

$$\tau_b = C_2\rho \left(w'^2 \right)$$

(12)

where $C_2 = 0.9$.

From discussions in Section 2.2.3, the velocity in Z direction plays a crucial role for the scouring process under ice cover. Hence, the above equation was used for calculation. The total bed shear stress in the scour hole can be expressed in dimensionless form as:

$$\hat{\tau} = \tau_b / \tau_0$$

(13)

where τ_0 is the bed shear stress of the approaching flow calculated from $\tau_0 = \rho U_{*c}^2$, where U_{*c} is the critical shear velocity.

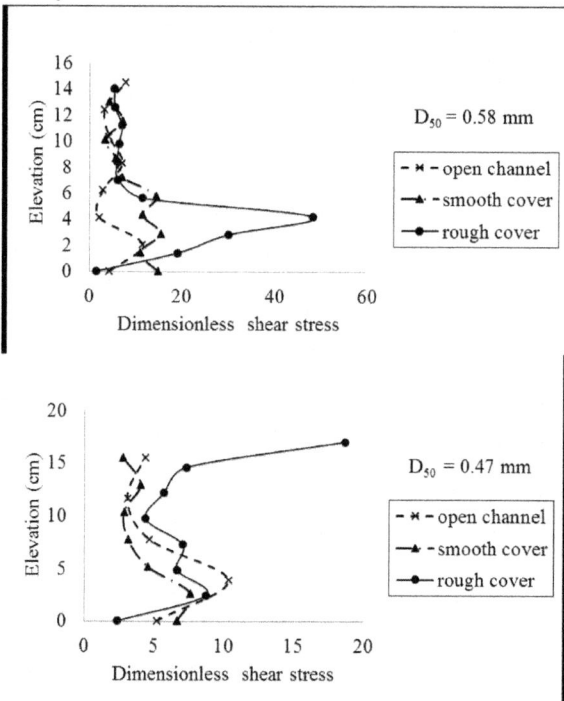

Figure 14. Comparison of measured dimensionless shear stress under different conditions.

Figure 14 shows the variation of dimensionless shear stress at different elevations around the square abutment in open channel flow, smooth ice cover, and rough ice cover. The following conclusions can be drawn:

- Under rough ice cover condition, the rate of change of dimensionless shear stress is the largest compared with smooth ice cover and open channel. A larger bed shear stress is needed for the incipient motion in the scour region.

- Finer particles need relatively less dimensionless shear stress for the scouring under the same flow and cover conditions.

- Under rough ice cover condition, the dimensionless bed shear stress at the underside of ice has a relatively large magnitude. Additionally, the variation of dimensionless shear stress under ice cover has a shape of "S" as shown in Figure 14. However, more research should be conducted to validate this.

CONCLUSION

Ice cover and nonuniform sediment are two critical parameters for the development of scour around hydraulic structures, such as bridge piers and abutments. By using two case studies, the impact of ice cover is presented in the chapter. With an increase in the densimetric Froude number, there is a corresponding increase in the dimensionless scour depth. For same nonuniform sediment, due to the formation of armor layer, less maximum scour depth was noted around bridge foundation structures. The increase in ice cover roughness can also result in a larger scour depth and profile. By analyzing the ADV data collected from Case 1, it was found that the velocity magnitude in Z direction is larger under ice covered flow in the scour hole than that under open channel flow. The velocity magnitude in Y direction has the largest magnitude under ice cover. More research is required to gain sufficient insight into the impact of ice cover on the local scour around hydraulic structures.

ACKNOWLEDGEMENTS

Experimental work was conducted at Quesnel River Research Center (QRRC), Likely, BC, and Sedimentation and Scour Study Laboratory at University of Windsor. Staff and colleagues provided great help.

REFERENCES

1.   Ackermann N L, Shen H T, Olsson P, 2002. Local scour around circular piers under ice covers. *Proceeding of the 16th IAHR Internnational Symposium on Ice*, Internnational Association of Hydraulic Engineering

Research, Dunedin, New Zealand.

2.   Acharya A, 2011. Experimental study and numerical simulation of flow and sediment transport around a series of spur dikes, PhD thesis, The University of Arizona, pp. 140-161.

3.   Biron P M, Robson C, Lapointe M F, Gaskin S J, 2004, Comparing different methods of bed shear stress estimates in simple and complex flow fields. *Earth Surf Process Landforms*, Vol. 29, pp. 1403-1415.

4.   Carey K L, 1966. Observed configuration and computed roughness of the underside of river ice, St Croix River, Wisconsin, Professional paper 550-B, US Geological Survey, pp. B192-B198.

5.   Chien N, Wan Z, 1999. *Mechanics of Sediment Transport*. ASCE Process. Reston, Virginia, USA.

6.   Dey S, Barbhuiya A K, 2005. Turbulent flow field in a scour hole at a semicircular abutment, Can J Civil Engin, Vol. 32, pp. 213-232.

7.   Duan J G, He L, Fu X, Wang Q, 2009, Mean flow and turbulence around experimental spur dike, *Adv Water Res*, Vol. 32, pp. 1717-1725.

8.   Ettema R, Braileanu F, Muste M, 2000. Method for estimating sediment transport in ice covered channels, *J Cold Region Engin*, *ASCE*, Vol. 14, No. 3, pp. 130-144.

9.   Ettema R, Daly S, 2004. Sediment transport under ice. ERDC/CRREL TR-04-20. Cold Regions Research and Engineering Laboratory U.S. Army Engineer Research and Development Center 72 Lyme Road Hanover, New Hampshire 03755.

10.   Goring D G, Nikora V I, 2002, Despiking acoustic Doppler velocimeter data. *J Hydraul Engin, ASCE*, Vol. 128, No. 1, pp. 117-126.

11.   Hager W H, 1999. *Wastewater Hydraulics: Theory and Practice*, Springer, Berlin, New York, pp. 17 – 54.

12.   Hains D B, 2004. An experimental study of ice effects on scour at bridge piers, PhD Dissertation, Lehigh University, Bethlehem, PA.

13.   Kim S C, Friedrichs C T, Maa J P Y, Wright L D, 2000. Estimating bottom stress in tidal boundary layer from acoustic Doppler velocimeter data, *J Hydraul Engin*, Vol. 126, No. 6, pp. 399-406.

14.   Kuhnle R A, Jia Y, Alonso C V, 2008. Measured and simulated flow near a submerged spur dike, *J Hydraul Engin, ASCE*, Vol. 134, No. 7, pp. 916-924.

15.   Lau Y L, Krishnappan B G, 1985. Sediment transport under ice cover, *J Hydraul Engin, ASCE*, 111(6), pp. 934-950.

16.   Li S S, 2012. Estimates of the Manning's coefficient for ice covered

rivers, *Water Management, Proc Institut Civil Engin*, Vol. 165, Issue WM9, pp. 495-505.

17. MacVirar B J, Roy A G, 2007. Hydrodynamics of a forced riffle pool in a gravel bed river: Mean velocity and turbulence intensity, *Water Res Res*, Vol. 43, Issue 12, DOI: 10.1029/2006WR005272.

18. Mays L W, 1999. *Hydraulic Design Handbook*, McGraw-Hill, pp. 3.12.

19. Melville B W, 1997. Pier and Abutment scour: integrated approach, *J Hydraul Engin, ASCE*, Vol. 123(2), pp. 125-136.

20. Meyer-Peter E, Müller R, 1948. Formula for bed-load transport, Proceedings of International Association for Hydraulic Research, 2nd Meeting, Delft, Netherlands, pp. 39-64.

21. Morse B, Hicks F, 2005. Advances in river ice hydrology 1999-2003, *Hydrol Process*, Vol. 19, Issue 1, pp. 247-263.

22. Munteanu A, 2004. Scouring around a cylindrical bridge pier under partially ice-covered flow condition, Master thesis, University of Ottawa, Ottawa, Ontario, Canada.

23. Munteanu A, Frenette R, 2010. Scouring around a cylindrical bridge pier under ice covered flow condition-experimental analysis, R V Anderson Associates Limited and Oxand report.

24. NCHRP Web-only Document 181, 2011. Evaluation of Bridge-Scour Research: Abutment and Contraction Scour Processes and Prediction. NCHRP Project 24-27(02).

25. Rehmel M, 2007. Application of Acoustic Doppler Velocimeter for streamflow measurements, *J Hydraul Engin, ASCE*, Vol. 133, Special Issue: Acoustic Velocimetry for Riverine Environments, pp. 1433–1438.

26. Sui J, Wang J, He Y, Krol F, 2010. Velocity profile and incipient motion of frazil particles under ice cover, *Int J Sedim Res*, Vol. 25(1), pp. 39-51.

27. Smith B T, Ettema R, 1997. Flow resistance in ice covered alluvial channels, *J Hydraul Engin ASCE*, Vol. 123(7), pp. 592-599.

28. Wahl T L, 2000, Analyzing data using WinADV, 2000 Joint Conference on water resources engineering and water resources planning and management, Minneapolis, Minnesota, pp. 1-10.

29. Wu P, Balachandar R, 2015. Measurement of scour profiles around bridge piers in channel flow with and without ice-cover, 22nd Canadian Hydrotechnical Conference, April 29-May 3, Montreal, Quebec, Canada, pp. 1-11.

30. Wu P, Hirshfield F, Sui J, 2014a, Armor layer analysis of local scour around bridge abutments under ice cover. *River Research and Applications*,

published online in Wiley Online Library, DOI: 10.1002/rra.2771.

31.   Wu P, Hirshfield F, Sui J. 2014b, Further studies of incipient motion and shear stress on local scour around bridge abutment under ice cover, *Can J Civil Engin*. Vol. 41(10), pp. 892-899.

32.   Wu P, Hirshfield F, Sui J, 2015. Local scour around bridge abutments under ice covered conditions – an experimental study, *Int J Sedim Res*, Vol. 30 (1), pp. 39-47.

33.   Yang C T, 2003. *Sediment Transport, Theory and Practice*. Krieger Publishing Company, Krieger Drive, Malabar, Florida 32950.

Chapter 3

ROLE OF HYDRAULIC CONDUCTIVITY UNCERTAINTIES IN MODELING WATER FLOW THROUGH FOREST WATERSHEDS

Marie-France Jutras and Paul A. Arp

Faculty of Forestry and Environmental Management, University of New Brunswick, Fredericton, Canada

INTRODUCTION

Soil hydraulic conductivities at saturation (K_{sat}) are highly variable in space and time. For example, K_{sat} varies along vertical and lateral flow paths depending on directional changes in soil texture, density, and structure [1, 2, 3]. Temporal changes are caused by changes in soil structure and bulk density (D_b) in response to, e.g., (i) gradual soil formation processes, and (ii) operationally induced soil compaction or de-compaction due to various land-uses [4]. Changes in weather and climate also affect K_{sat} through freezing and thawing [5, 6], swelling and shrinking [7], extent of rooting and related organic matter build-up [8]. This chapter explores how changes in hydraulic conductivity may affect modelled rates of water flow through forested watersheds, with flows referring to infiltration, percolation, run-off, interflow, base flow, and stream discharge. This is done by way of sensitivity analyses centered on two well-studied watershed studies, referring to Moosepit Brook, Nova Scotia [1,9] and Turkey Lakes, Ontario [1, 10]. Also addressed are:

K_{sat} impacts on the retention of soil water and the transmittance of the same towards streams as influenced by evapotranspiration from open conditions to forests [14]; the relationship between K_{sat} and the state of organic matter decomposition, as characterized by the von Post index from fibric (H1) to fully humified or sapric (H10) [11, 12, 13].

The sensitivity analysis is based on using the forest hydrology model ForHyM2 [1, 15] to determine how scenario-set K_{sat} variations affect soil water retention and flow including stream discharge through the watersheds. The scenarios vary K_{sat} by changing organic matter (OM) and sand content from their actual values within the 0 to 100% per soil weight range.

Quantitative Background

The equations used for estimating the sensitivity of K_{sat} on account of changes in soil texture, structure, density and organic matter content is given by [16], as follows:

$$\log_{10} K_{sat} = a + 7.94\log_{10}(D_p - D_b) + 1.96\ \text{SAND} \tag{1}$$

$$D_b = \frac{1.23 + (D_p - 1.23 - 0.75\ \text{SAND})(1 - \exp(-0.0106\ \text{DEPTH})}{1 + 6.83\ \text{OM}} \tag{2}$$

$$\frac{1}{Dp} = \frac{\text{OM}}{Dp_{om}} + \frac{1 - \text{OM}}{Dp_{min}} \tag{3}$$

where, Dp_{om} is the particle density of OM (1.3 gcm⁻³), Dp_{min} is the particle density of mineral soils (2.65gcm⁻³), SAND and OM are dry soil weight fractions (fine earth fraction only), DEPTH is the mid depth of each soil layer (cm), "a" represents K_{sat} when $D_p - D_b = 1\ \text{g cm}^3$ and SAND = 0%. Fig. 1 illustrates how variations in D_b, OM, and SAND affect K_{sat} in general.

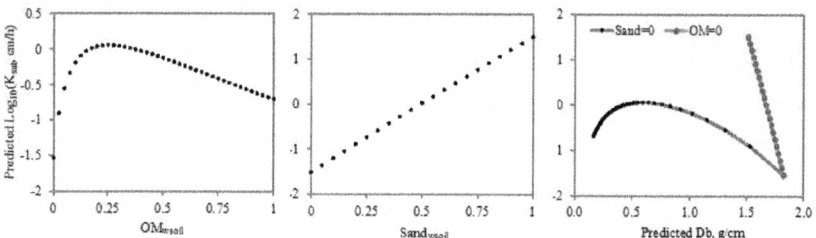

Figure 1. Left and middle: how $\log_{10} K_{sat}$ varies with increasing OM, and sand fraction. Right: Changes in $\log_{10} K_{sat}$ and Db when OM and Sand fraction = 0.

For organic soils, it is important to adjust a, D_b, and D_p in Eqs. 1 to 3 by extent of organic matter decomposition and humification [12, 11, 17, 13, 18, 19]. These adjustments are based on the von Post humification index ([11, 20], Table 1) as follows:

$$D_b = 0.035 + 0.0159\ vP\ (R2 = 0.93) \tag{4}$$

$$D_{p\,om} = \frac{D_b}{1 - \phi} \tag{5}$$

$$\phi = 100.38 - 76.7D_b \ (R^2 = 0.99) \tag{6}$$

$$a = (2.05 \pm 0.2) - (0.046 \pm 0.004)vP^2 \ (R^2 = 0.89) \tag{7}$$

where vP is the von post index (Table 2) and ϕ is the soil porosity. Fig. 2 illustrates the relationship between the von Post adjusted $\log_{10}K_{sat}$ (Eqs. 1, 4-7) and $\log_{10}K_{sat}$ based on literature sources. Fig. 2shows (i) a plot of actual versus best-fitted K_{sat} values (left), and actual as well as best-fitted K_{sat} values with increasing organic matter humification in peaty soils (right).

Table 1. von Post humification index, with K_{sat}, D_b and D_p for 100% OM content according to Eqs. 1 and 4to 7; adapted from [21] and [11]

Peat Class	von Post Index	Squeeze Test: Exudate condition	$\log_{10Ksa}t$	K_{sat} [1] cm^{h-1}	D_b [2] g cm^{-1}	D_p [2] g cm^{-1}
Fibric	H1	Water colourless	3.15	1406.48	0.05	1.44
Decomposition: none to slight; Amorphous content: low	H2	Water yellowish	2.88	756.62	0.07	1.41
	H3	Water brown, muddy; no peat	2.55	356.24	0.08	1.39
Mesic	H4	Water dark brown, muddy; no peat	2.15	141.08	0.10	1.37
Fibers still recognisable;	H5	Water muddy; some peat	1.66	46.20	0.11	1.36
Decomposition: moderate to strong; Amorphous content: medium	H6	Water dark brown; 33% peat	1.09	12.40	0.13	1.36
	H7	Any water very dark brown, 50% peat	0.43	2.72	0.15	1.35
Sapric	H8	66% peat, Water pasty	-0.32	0.48	0.16	1.34
Fibers unrecognisable;	H9	Nearly all peat; paste uniform	-1.16	0.07	0.18	1.34
Decomposition: very strong to complete; Amorphous content: high	H10	100% peat paste; no water	-2.09	0.006	0.19	1.34

[i] - [1]Eq. 1 from [16]

[ii] - [2]Eqs. 4-5 from [12]

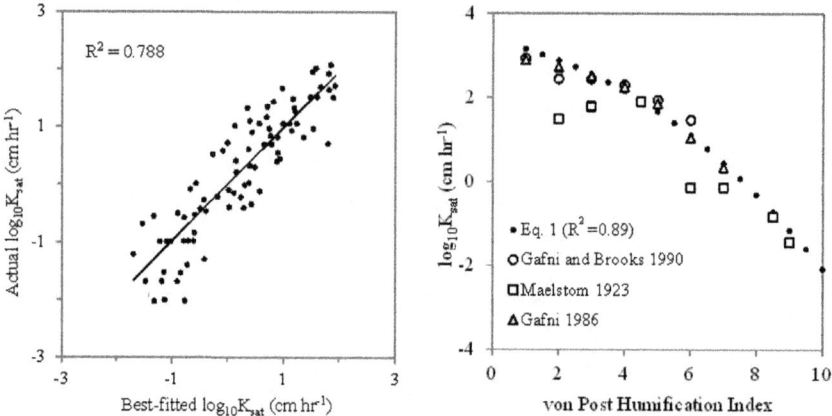

Figure 2. Best-fitted $\log_{10}K_{sat}$ versus actual data from New Brunswick and Nova Scotai, Canada, as seen in [1] and [16] (left), best-fitted $\log_{10}K_{sat}$ versus von Post humification index from literature sources (right).

METHODS

The two study areas, Moosepit Brook and Turkey Lakes, have contrasting terrain (generally flat versus hummocky), climate (maritime versus continental), vegetation (mostly coniferous versus deciduous), and soil parent material (ablation till versus basal till) (Table 2, Fig. 3)

Eight scenarios were adopted to examine the impacts of K_{sat} variations on water flow through these locations, as follows: the actual soil conditions in terms of soil texture and organic matter (Scenario 1, Table 3), varying the soil texture sand, silt, or clay (Scenarios 2, 3, and 4), and varying the soil organic matter content (Scenarios, 5, 6, 7, and 8).

Scenario 1:

Actual soil texture and OM content

Scenarios 2 to 4: Changing soil texture

- Sand = 95% sand, 1% silt, 4% clay
- Silt = 7% sand, 87% silt, 6% clay
- Heavy clay = 25% sand, 25% silt, 50% clay

Scenarios 5 to 8: Changing organic matter content

- Half actual OM
- Double actual OM
- No OM throughout entire soil profile
- 100% OM throughout entire soil profile using the von Post profile of FF = 3, A&B = 5, C = 9, Subsoil = 10.

Table 2. Site description for the Moosepit Brook and Turkey Lakes watersheds.

Watershed characteristics	Moosepit Brook	Turkey Lakes
	Nova Scotia (NS)	Ontario (ON)
Abbreviation	MP	TL
Latitude (N)	44°28'	47°03'
Longitude (W)	65°03'	84°25'
Area (ha)	1670	1050
Elevation (m)	100-150	350-400
Slope (%)	1	8
Deciduous:coniferous	50:50	100:0
Rooting habit	Medium	Deep
Forest floor depth (cm)	5	7
Mineral soil: depth (cm); texture	50; SL	60; SilL
Subsoil: depth (cm); texture	70; LS	100; LS
Bedrock	Metamorphic greenschist slate	Metavolcanic basalt
Land Formation	Glacial till	Ablation till on basal till
Topography	Rolling	Undulating to rolling
Mean yearly temperature (°C)	7.02	4.52
Mean yearly snow depth (cm)	5	23
Mean yearly rainfall (mm)	1140	790
Model Run Years	1999-2004	1997-2004

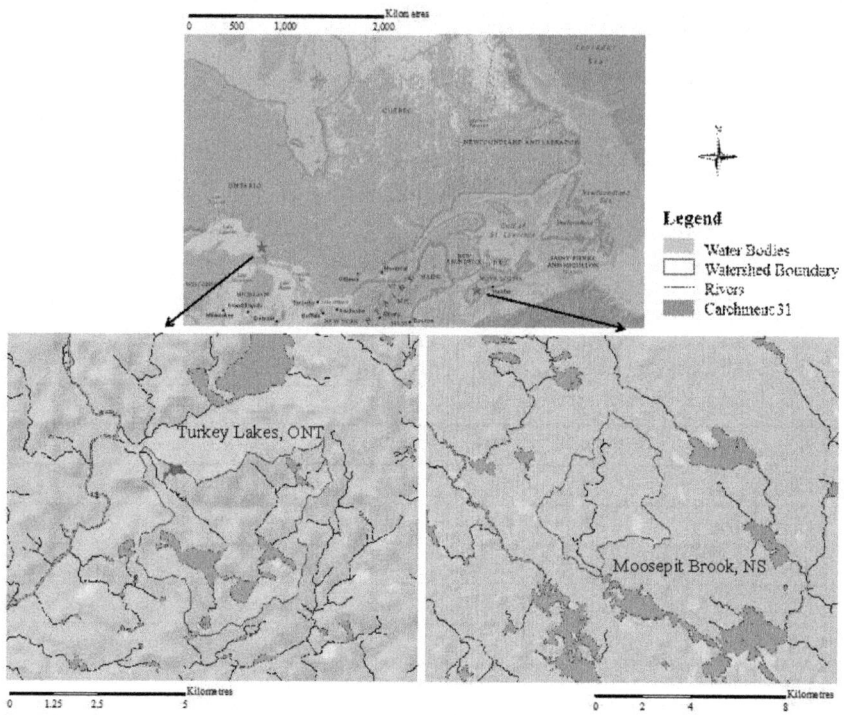

Figure 3. Locator maps for the Turkey Lakes (left) and Moosepit Brook (right) study areas.

Actual watershed inputs: Scenario 1

Table 3. Actual scenario soil input for Moosepit Brook and Turkey Lakes.

Horizon	Moosepit Brook, NS					Turkey Lakes, ONT				
	Depth (cm)	Sand (%)	Silt (%)	Clay (%)	OM (%)	Depth (cm)	Sand (%)	Silt (%)	Clay (%)	OM (%)
A	21	66	24	10	4.5	15	66	24	10	5
B	21	66	24	10	9	30	66	24	10	3
C	50+	82	12	6	4.5	50+	22	65	13	2

The sand texture percentages for scenarios 2-4 demonstrate the effects of varying texture on K_{sat} from sandy and sandy loam soils to silty and clayey soils (Fig. 4). The organic matter levels for scenarios 5-8 were chosen to demonstrate the effects of changing the organic matter on content from very small in mineral soils to fully organic soils. For the 100% organic soil condition (Scenario 8), three sub-scenarios were chosen to account for variations in forest cover from 100 % (fully forested), 50% (varying from forested to boggy) and 0% (open moss and shrub-covered bogs with no trees). This is to demonstrate how varying K_{sat} levels from high to low increase the amount of water available for evapotranspiration

Each scenario was used for initializing the ForHyM2 requirements for soil texture and organic matter by soil layer, with the A and B layers representing the top soil conditions, and the C layer representing the subsoil conditions Table 2). Layer-specific values for D_p, D_b and K_{sat} were then generated automatically via Eqs. 1 to 3. All other site-specific input requirements for daily weather (rain, snow air temperature), slope, aspect, elevation and soil layer depths were kept the same. Scenario 1 was used to refine the Eq. 1 estimates for K_{sat}, by adjusting the K_{sat} adjustment multipliers for surface run-off, interflow (forest floor, A&B layers combined), baseflow (C layers combined), infiltration, and soil percolation from the forest floor to the topsoil, and from the topsoil to the subsoil. The calibrations were done by matching modeled with actual stream discharge a the daily level, using local weather records for daily rain, snow and air temperature as model input. Modelled snowpack depth was also calibrated using daily snowpack data. The ForHyM2 model runs were done for 1999 – 2004 for Moosepit Brook, and for 1997-2002 for Turkey Lakes.

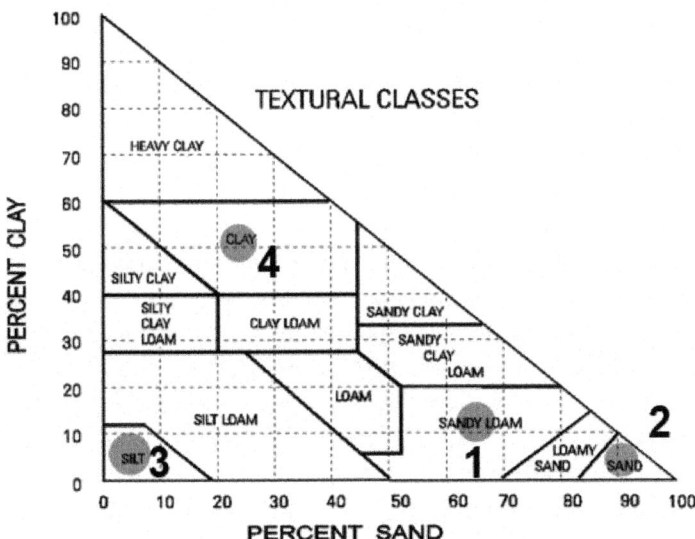

Figure 4. Mineral texture class triangle for fine soil showing texture classes for scenarios 1 - 4 (adapted from CANSIS 2000).

RESULTS

The results of this analysis are shown in Tables 3 to 8 and in Figs. 5 to 12 for the Moosepit Brook and Turkey Lakes study areas. Tables 3 and 4 inform about the Scenario-based changes on Dp, Db and K_{sat} for each of the two sites by topsoil and subsoil. The Db numbers indicate that the subsoil at both locations is compacted, with K_{sat} values typically 10 to 50 times lower in the subsoil than in the topsoil. Since the soil texture is sandier at Moosepit Brook than at Turkey Lakes, K_{sat} values remain higher in the subsoil at Moosepit Brook than at Turkey Lakes. Changing the topsoil texture from the actual values changes K_{sat} by about 5x upwards, and by about 10x downwards at both locations. These K_{sat} changes are similar for the somewhat coarser subsoil at Moosepit Brook. In contrast, subsoil K_{sat} is not much affected by increasing the clay and silt content, but increases with increasing sand content towards 95% by a factor of 147

Table 5 and Fig. 5 inform about the 5-year cumulative effects of the texture and OM changes on ForHyM2-modelled run-off, forest floor interflow, topsoil interflow, baseflow and stream discharge in terms of modelled mm per study period, and also in terms of modeled flow rate percentages per stream discharge. As shown, the interflow and baseflow percentage contributions to stream discharge so compiled are very sensitive to K_{sat} as well as basin slope: for intermediate K_{sat} values, interflow would dominate the base flow contributions to stream discharge within the steeper watershed at Turkey Lakes (average slope = 8%). The reverse would occur at the flatter Moosepit Brook watershed (average slope = 1%). Low subsoil permeability at Turkey Lakes would further accentuate this difference. In detail, base flow would dominate in both watersheds or at any location within the watersheds with high soil permeability and where the subsoil would not be blocked by impervious bedrock. In contrasts, locations with low overall soil permeability and low slopes would be most variable in terms of their cumulative run-off, interflow and baseflow contributions, varying from mostly baseflow to mostly interflow (Fig. 6). For example, mineral soils with high silt content (Scenario 3) would support more lateral flow in the topsoil as opposed to soils with high sand content (Scenario 2). Doubling the OM in the mineral soil (Scenario 5) would also increase baseflow, whereas reducing OM (Scenario 6) would induce the opposite. The extent of water infiltration in Scenario 4, as modeled, would be midway between Scenarios 2 and 3

Table 4. Results for various levels of sand and OM against K_{sat}, D_b and D_p for Moosepit Brook and Turkey Lakes.

Site	Scenarios	K_{sat}, cm $^{h-1}$		D_b, g c^{m-1}		D_p, g c^{m-1}	
		Mineral	Subsoil	Mineral	Subsoil	Mineral	Subsoil
Moosepit Brook	1: Actual	48.40	5.95	0.95	1.61	2.48	2.59
	2: Sand	162.90	29.15	0.93	1.50	2.48	2.59
	3: Silt	3.05	0.15	1.00	1.86	2.48	2.59
	4: Heavy clay	7.15	0.50	0.99	1.80	2.48	2.59
	5: Double OM	60.60	13.30	0.72	1.48	2.33	2.54
	6: Half OM	31.35	2.70	1.14	1.70	2.56	2.62
	7: No OM	12.60	0.75	1.41	1.83	2.65	2.65
Turkey Lakes	1: Actual	39.80	0.10	1.09	1.85	2.55	2.61
	2: Sand	136.25	14.70	1.06	1.54	2.55	2.61
	3: Silt	2.50	0.05	1.15	1.92	2.55	2.61
	4: Heavy clay	5.80	0.10	1.15	1.84	2.55	2.61
	5: Double OM	56.05	0.30	0.90	1.68	2.45	2.57
	6: Half OM	30.55	0.05	1.19	1.85	2.59	2.61
	7: No OM	14.50	0.00	1.39	2.06	2.65	2.65

Figs. 6 to 9 inform about the changes in daily variations in run-off, interflow and baseflow for both locations as the soil texture changes from actual to sandy, silty and clayey (Scenarios 1 to 4, respectively, Figs. 6, 7), and soil organic matter content changes actual to 0.5 and 2 x, and 100% (Scenarios 1, and 5 to 8, Figs. 8, 9). As shown, these flows would peak faster with increasing K_{sat} (increasing sand and organic matter content), and would saturate the lower soil layers more quickly with decreasing K_{sat} and decreasing pore space, or increasing bulk density. Among the scenarios, the largest textural change on the flow regime was incurred by increasing the silt content within the already compacted subsoil at Moosepit Brook. Note that organic soils with 100% sapric organic matter would also have very low interflow and baseflow rates, and would therefore lead to relative fast soil saturation as well.

Table 5. Lateral stream discharge by cumulative and percent runoff, interflow, and base flow for scenarios 1-8 for Moosepit Brook (1999-2004) and Turkey Lakes (1997-2004).

Site	Scenario	Runoff (mm)	%2	Interflow FF (mm)	%2	Interflow A&B (mm)	%2	Base flow (mm)	%2	Total Discharge (mm)
Moosepit Brook	1	2.7	0.1	202.3	4.8	682.8	16.1	3341.2	79.0	4229.0
	2	2.6	0.1	202.4	4.7	512.0	11.9	3580.4	83.3	4297.0
	3	7.0	0.2	209.5	5.1	2653.0	64.4	1251.5	30.4	4121.0
	4	2.7	0.1	202.4	4.8	1609.7	38.3	2391.4	56.9	4206.0
	5	2.6	0.1	202.3	4.8	450.0	10.6	3578.0	84.5	4233.0
	6	2.7	0.1	202.4	4.8	964.0	22.8	3054.8	72.3	4224.0
	7	2.7	0.1	202.3	4.8	1675.8	39.8	2329.9	55.3	4210.7
	8	0.0	0.0	594.8	13.2	3771.1	83.4	153.7	3.4	4519.6
	8[1]	0.0	0.0	612.4	11.6	4495.9	85.3	161.0	3.1	5269.3
	8[2]	0.0	0.0	668.6	10.6	5487.3	86.7	171.1	2.7	6327.0
Turkey Lakes	1	1.1	0.0	375.9	9.1	2990.0	72.2	772.9	18.7	4139.0
	2	3.3	0.1	376.1	8.8	301.0	7.1	3588.5	84.1	4269.0
	3	114.9	2.8	521.2	12.8	2635.0	64.8	796.5	19.6	4068.0
	4	19.4	0.5	389.1	9.4	1965.0	47.3	1782.1	42.9	4156.0
	5	2.2	0.1	376.1	9.1	2314.0	55.8	1454.4	35.1	4146.0
	6	0.5	0.0	375.8	9.1	3273.0	79.2	485.6	11.7	4135.0
	7	0.0	0.0	374.8	9.1	3457.0	83.9	290.7	7.1	4122.0
	8	0.0	0.0	69.8	1.5	4250.5	92.1	292.9	6.3	4613.1
	8[1]	0.0	0.0	70.9	1.2	5463.3	93.6	305.0	5.2	5839.2
	8[2]	0.0	0.0	72.6	1.0	6609.0	94.6	305.0	4.4	6986.5

[1] 50% coverage [2] 10% coverage [2] % values refer to the calculated percent contributions of run-off, FF interflow A&B interflow and baseflow to stream discharge.

Assessing the waterflow through peatland locations within each of the two watersheds, and setting the state of decomposition of the peat equal to H1, H4, H7 and H10 produced the results compiled in Table 6. As shown, organic soils mostly composed of fibric to mesic peat (H1) would support deep percolation and baseflow, whereas organic soils mostly composed of humic peat (H10) would contain pooled water from the subsoil upwards to the surface, thereby encouraging surface run-off

Note also from Table 4 and 6 that the changing K_{sat} values for decomposing peat would also have strong effects on forested peatland evapotranspiration and on stream discharge: the lower K_{sat}, the higher would be the rate of water retention and subsequent forest water uptake and evapotranspiration during the growing season (Fig. 10). In contrast, the higher K_{sat}, the faster water would be lost due to quick baseflow (Fig. 11). Outside the growing season, run-off increases, as modeled and as to be expected (Fig. 10 and 11)

Table 6. Lateral stream discharge by cumulative and percent runoff, interflow, and baseflow for scenario 8 to represent 100% peat surface deposits (von Post index set at H1, H4, H7, and H10 for the entire profile) underneath forest cover at each of the two locations.

Site	von Post	Runoff (mm)	%	Inter-flow FF (mm)	%	Inter-flow A&B (mm)	%	Base flow (mm)	%	Total Dis-charge (mm)
Moosepit Brook	H1	0.00	0.0	1.09	0.0	53.00	1.1	4618.27	98.8	4672.35
	H4	0.00	0.0	20.27	0.4	222.67	4.9	4291.43	94.6	4534.38
	H7	0.00	0.0	240.34	5.9	184.86	4.5	3676.06	89.6	4101.26
	H10	3275.13	97.3	74.12	2.2	13.32	0.4	4.67	0.1	3367.25
Turkey Lakes	H1	0.00	0.0	0.08	0.0	0.84	0.0	5954.36	100.0	5955.27
	H4	0.00	0.0	1.03	0.0	7.64	0.1	5641.93	99.8	5650.61
	H7	0.00	0.0	55.87	1.2	8.18	0.2	4655.45	98.6	4719.50
	H10	3057.12	97.7	50.55	1.6	4.51	0.1	16.86	0.5	3129.03

The extent water retention in terms of mm per soil layer is illustrated in Fig. 12 for the two study locations as modeled for the actual soil (Scenario 1) and for organic soil conditions (100% organic matter content, Scenario 8), starting the soil moisture content at field capacity for January 1. For the slowly draining peatland scenario (Scenario 8), subsoil moisture levels would increase from field capacity towards saturation in about one year. For the well-drained upland soil conditions (Scenario 1), K_{sat} values would be sufficiently high so that soil moisture conditions would fluctuate around the field capacity, depending on season as well as rainfall and snow melt events

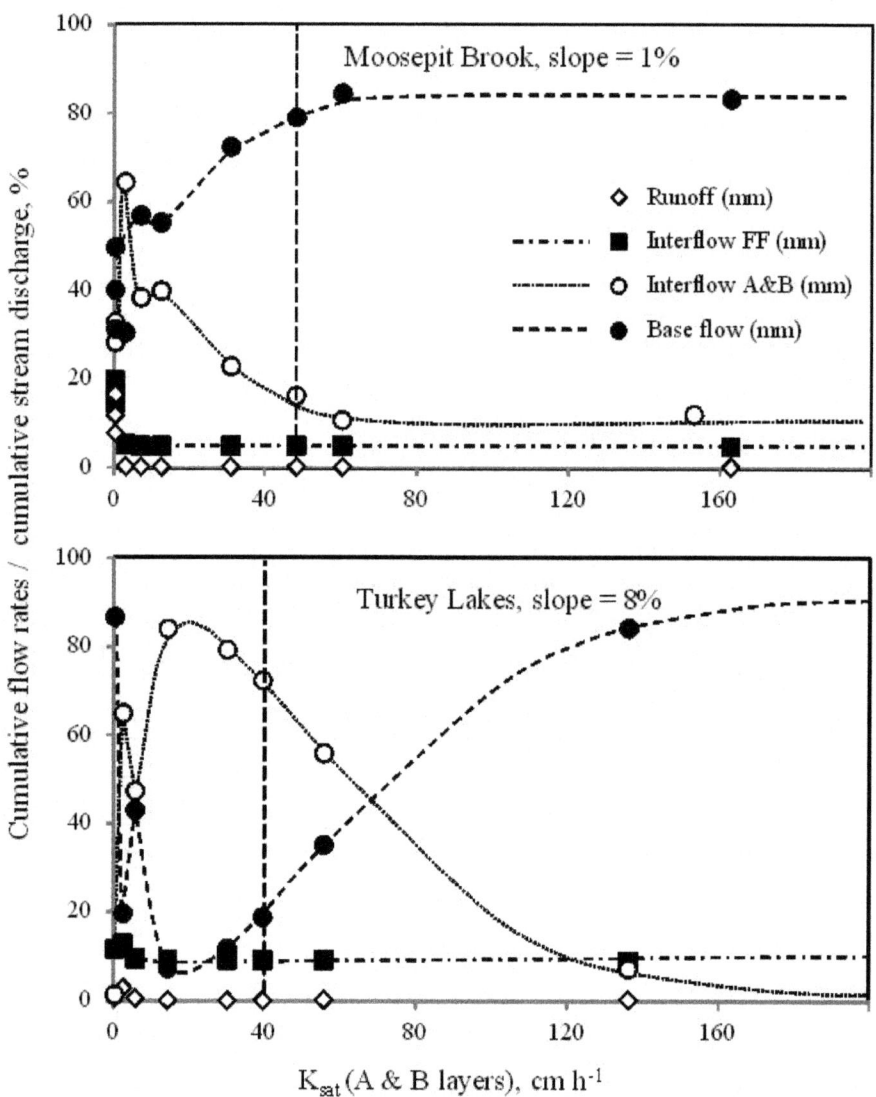

Figure 5. K_{sat} of the A&B layers by cumulative stream discharge % for Moosepit Brook (top), and Turkey Lakes (bottom) across all 8 scenarios (vertical dashed line represents the actual scenario).

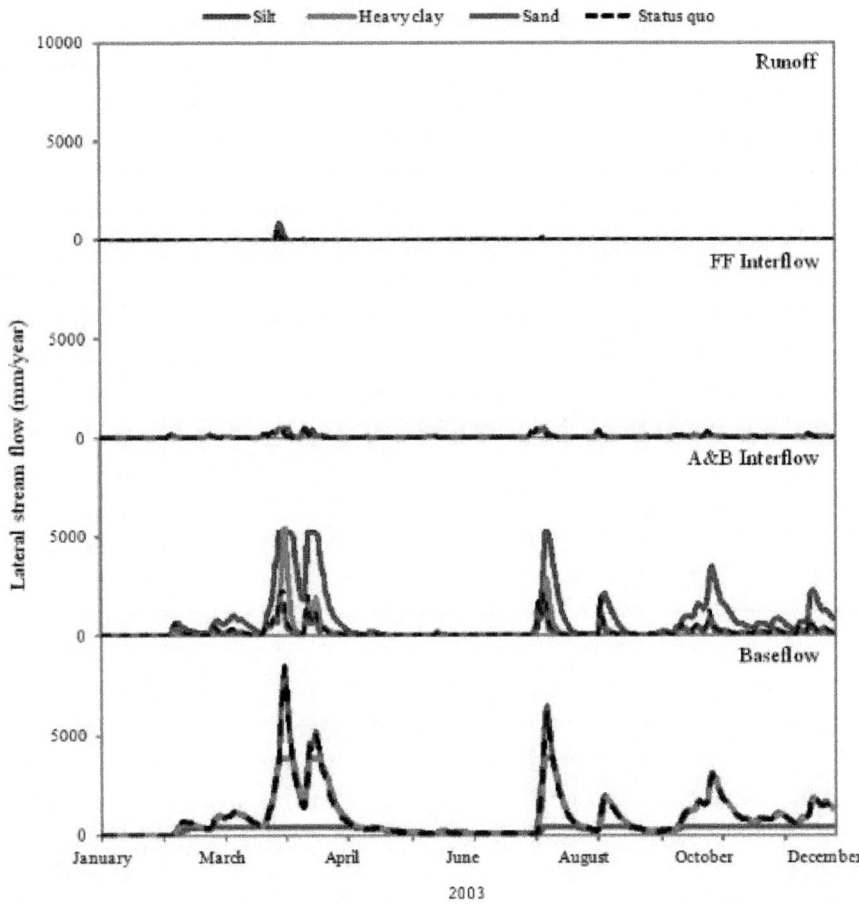

Figure 6. Run-off, forest floor and A&B interflow and base flow for Moosepit Brook, by scenario from actual to sandy, silty and clayey (Scenarios 1 to 4, respectively; 2003)

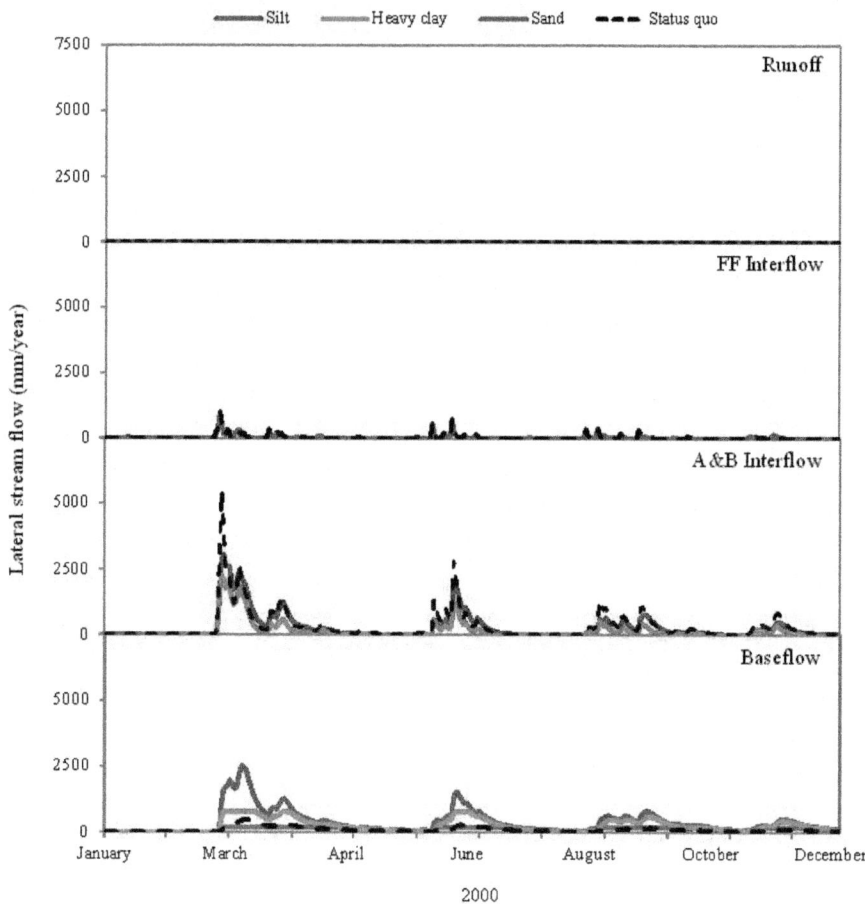

Figure 7. Run-off, forest floor and A&B interflow and base flow for Turkey lakes, by scenario from actual to sandy, silty and clayey (Scenarios 1 to 4, respectively; 2000), no runoff for any of the scenarios.

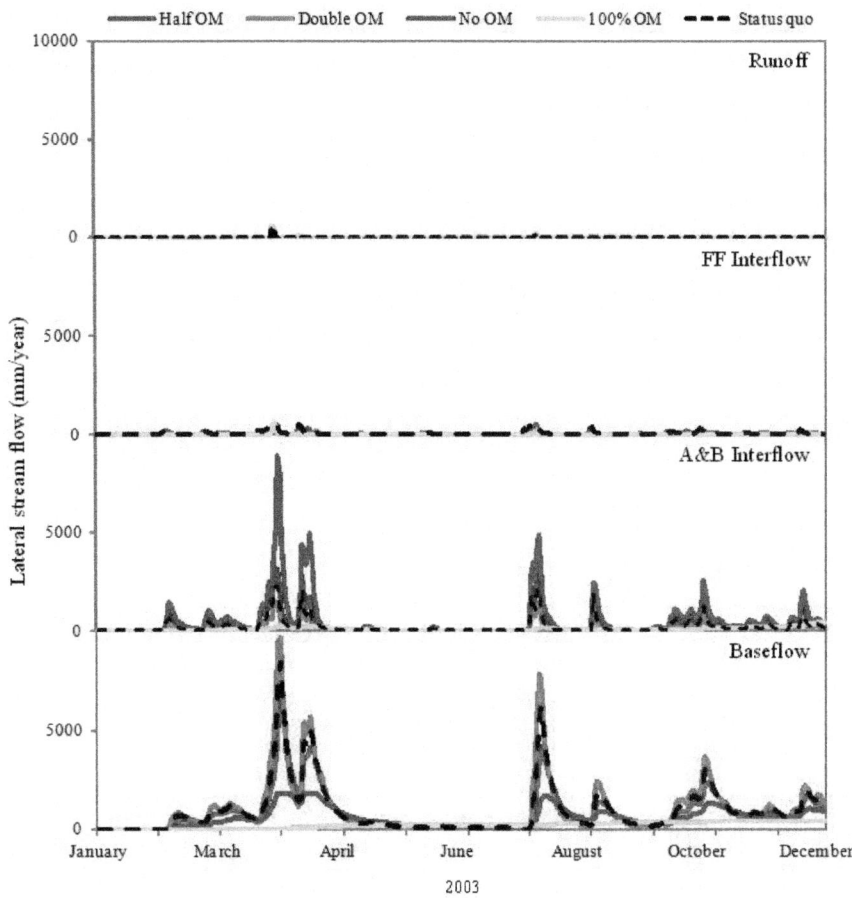

Figure 8. Run-off, forest floor and A&B interflow and base flow for Moosepit Brook, by scenario from actual to no, 0.5x and 2x actual organic matter content, and 100 % sapric organic matter (Scenarios 1 and 5 to 8, respectively; 2003).

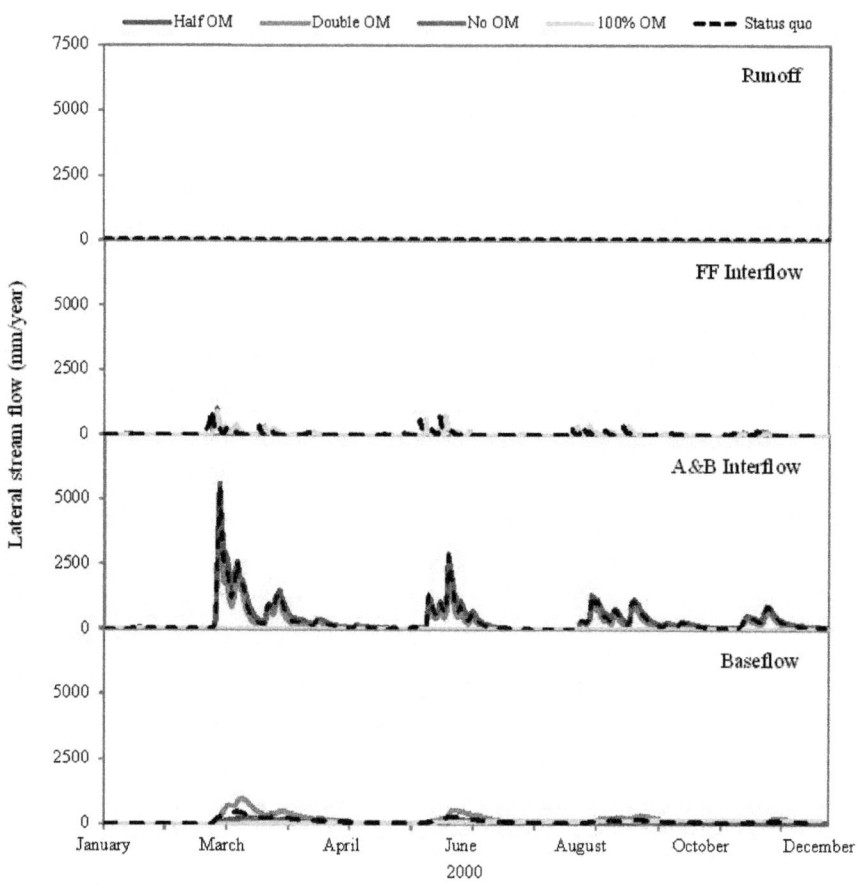

Figure 9. Run-off, forest floor and A&B interflow and base flow for Turkey Lakes, by scenario from actual to no, 0.5x and 2x actual organic matter content, and 100% sapric organic matter (Scenarios 1 and 5 to 8, respectively; 2000), no runoff for any of the scenarios.

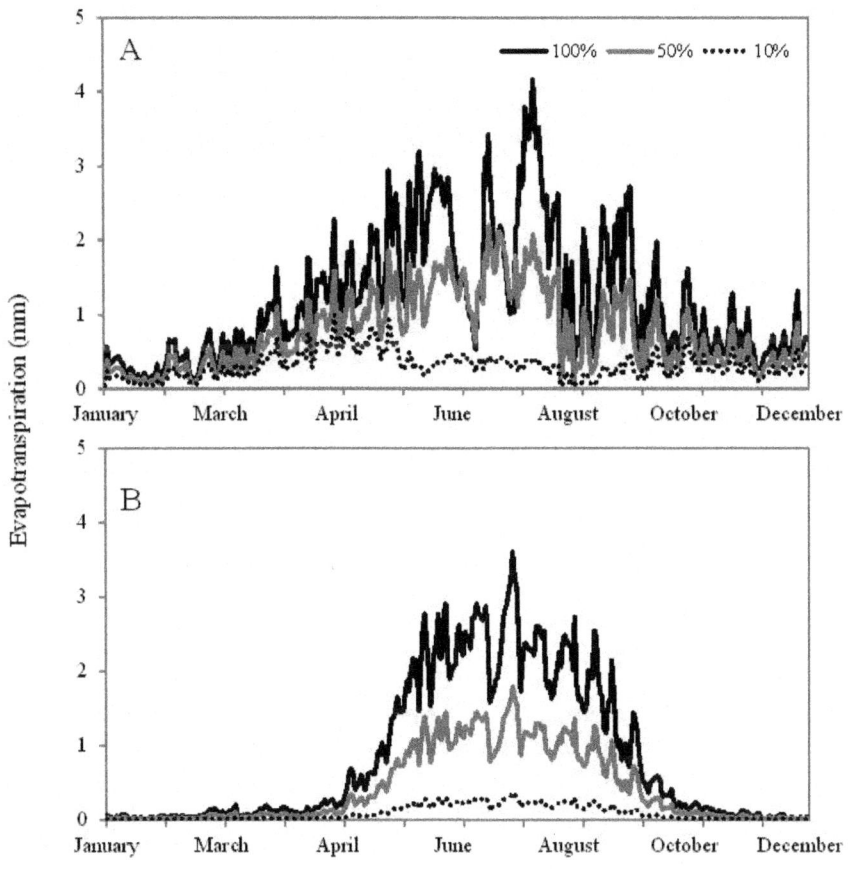

Figure 10. Evapotranspiration at Moosepit Brook (A) during 2003 and Turkey Lakes (B) during 2000, for Scenario 8 (100% OM), with actual (100% vegetation), Scenario 8[1] (50% vegetation), and Scenario 8[2] (10% vegetation). Note the difference in the extent of the growing season: wide for Moosepit Brook (maritime climate), and narrow for Turkey Lakes (continental climate).

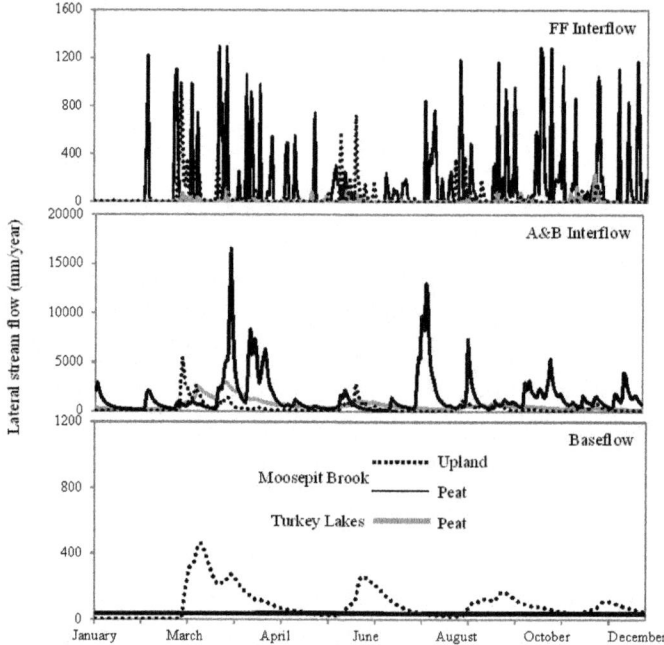

Figure 11. ForHyM2 estimated rates for daily forest floor and A&B interflow and base flow for peatland locations with a fibric - mesic – sapric layer profile at Moosepit Brook and Turkey Lakes, with 100% forest cover. Also shown: upland interflows and baseflows for the Moosepit Brook watershed (2000).

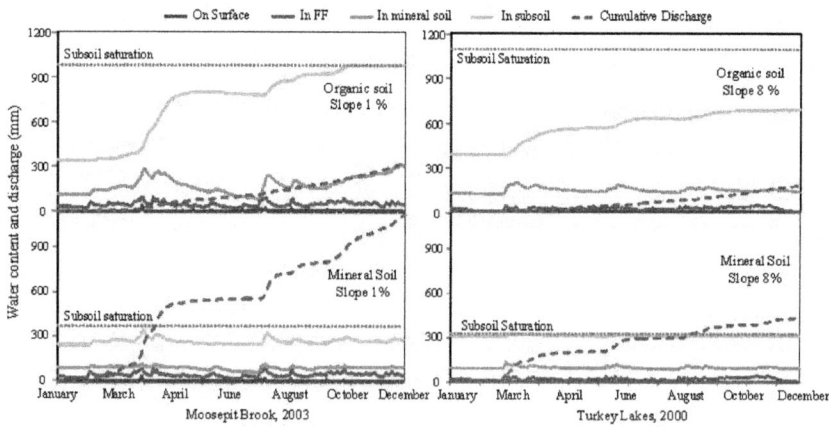

Figure 12. Soil water content on the surface, in the forest floor, in the mineral soil, and in the subsoil, as well as the cumulative discharge for Scenario 8 regarding organic

soil (100% OM, top), and actual mineral soil conditions (bottom), for Moosepit Brook (2003, left) and Turkey Lakes (2000, right). Simulations start with unsaturated soil condition. Discussion.

The above watershed-based K_{sat} evaluations have shown that the effective K_{sat} values for downward and lateral flow generally vary by a factor of 2 to 3 in comparison to corresponding values generated viaEqs. 1-3 [1]. As illustrated via Table 5 and subsequent figures, these variations lead to uncertainties in quantifying how water percolates through watersheds as run-off, interflow and baseflow (Fig. 5). These uncertainties also affect the flow response time, ranging generally from small delays to extended periods of flow as K_{sat} values decrease (Figs. 6 to 9). Across watersheds, however, flows tend to be well synchronized, regardless of major differences in texture, density, and organic matter content [28]. Typically, watersheds with the more compacted soils and therefore low K_{sat} values will be more peaked and will therefore be flashier than watersheds that allow deep percolation [25, 26, 27, 2]. The strongest impact of shallow to deep flow would deal with the water quality: deep water percolation during summer would lead to cooler and purer stream and seepage water with elevated pH than shallow water percolation [6]. During winter, deep percolation and persistent base flow would be warmer compared to the frost-affected surface water on poorly drained soils [6, 5]. Water flowing along the surface would also be more colored towards brown and more acidic than the more filtered and mineral-exposed water flowing at greater soil and subsoil depth [28]

While organic matter and soil density would not change drastically throughout undisturbed watersheds, such changes would occur during and after times of intense surface operations, especially under poor weather conditions. For example, forest operations during times of poor soil trafficability lead to ruts and increased soil compaction [29, 30]. In turn, soil compaction leads to lower K_{sat} values and therefore lower infiltration and hence higher surface run-off rates, thereby accelerating soil erosion and subsequent sediment transfer to streams and lakes [4]. Trails across the slopes of watersheds also affect downslope flow by compacting the soil underneath the trails, which means more water retention upslope along the trails, therefore leading to weather-effected trail destabilization, unless ditches and cross drains are installed to divert the water away from the trail beds [31]. Changes in forest cover could lead to changes in rooting space, which would – in turn – reduce the organic matter content within top and subsoils. This reduction would then alter the overall interplay between surface runoff, interflow and baseflow. Similarly, variations in climate from wet to dry (induces soil shrinking, may reduce root biomass), from frozen to non-frozen (induces collapse of frozen soil structures) would also affect K_{sat} and flow through soils by affecting the organic matter build-up, the state of soil organic matter humification, and overall changes in granular, blocky and columnar soil structures

The main advantage of the above K_{sat} formulation is that it allows for daily weather-related projections concerning downward and lateral water flow rates in forested to non-forested watersheds from times when soils are at saturation to times when soils are dry. At times of soil saturation, this quantification can then be used to estimate the effects of flow on soil stability and stream discharge. At times of drought, this quantification is can be used to estimate the effects of no flow on the remaining water reserves within soils and watersheds with and without peatland components (Fig. 12). Using ForHyM2 has the additional advantage of conducting these calculations year-round, summers through winters, based on already existing daily weather records, and extending these by way of daily, weekly, monthly or annual weather forecasts

ACKNOWLEDGEMENTS

Financial support for this research was received from Environment Canada, Alberta Sustainable Resource Department, and NSERC (Discovery Grant and CRD project grants).

REFERENCES

1. Jutras MF, Arp PA. Determination of hydraulic conductivity from soil characteristics and its application for modelling stream discharge in forest catchments. In (Ed.) LE, editor. Hydraulic Conductivity - Issues, Determination and Applications.: InTech; 2011. 189-202.

2. Brady NC, Weil RR. The Nature and Poperties of Soils. 13th ed. Helba S, editor. Upper Saddle River: Prentice Hall; 2001.

3. Crawford JW. The relationship between structure and the hydraulic conductivity of soil. European J Soil Sci. 1994;45: 493-502.

4. Harr RD, Fredriksen RL, Rothacher J. Changes in streamflow following timber harvest in southwestern Oregon. Forest Service Resoure Paper PNW-249. Portland, Oregon: Pacific Northwest Research Station, USDA; 1979.

5. Chi X. Hydrogeological assessment of stream water in forested watershed: temperature, dissolved oxygen, pH, and electrical conductivity. MSc. Thesis. Fredericton: University of New Brunswick, Forestry; 2008.

6. Steeves MT. Pre- and post-harvest groundwater temperatures and levels, in upland forest catchments in Northern New Brunswick. MSc. F Thesis.

Frederiction, NB: University of New Brunswick, Department of Forestry; 2004.

7. Bouma J. Field measurement of soil hydraulic properties characterizing water movement through swelling clay soils. J Hydol.. 1980; 45: 149-158.

8. Löfkvist J. Modifying soil structure using plant roots. PhD Thesis. Uppsala: Swedish University of Agricultural Sciences, Department of Soil Science; 2005.

9. Yanni S, Keys K, Meng FR, Yin X, Clair T, Arp PA. Modelling hydrological conditions in the maritime forest region of south-west Nova Scotia. Hydrol. Processes. 2000b; 14: 195-214.

10. Arp PA, Yin X. Predicting water fluxes through forests from monthly precipitation and mean monthly air temperature records. Can. J. For. Research. 1992; 22: 864-877.

11. Verry ES, Boelter DH, Paivanen J, Nichols DS, Malterer T, Gafni A. Physical properties of organic soils. Chapter 5. In Kolka RK, Sebestyenm SD, Verry ESBKN, editors. Peatland biogeochemistry and watershed hydrology at the Marcell Experimental Forest. Boca Raton: CRC Press; 2011. 135-176.

12. Päivänen J. Hydraulic conductivity and water retention in peat soils. Acta Forestalia Fennica. 1973; 129: 1-70.

13. Letts MG, Roulet NT, Comer NT, Skarupa MR, Verseghy DL. Parametrization of peatland hydraulic properties for the Canadian land surface scheme. Atmosphere-Ocean. 2000; 38: 141-160.

14. Boelter DH. Important physical properties of peat materials. In Proceedings, third international peat congress; 1968 August 18-23. Quebec, Canada: Department of Energy, Minds and Resources and National Research Council of Canada; 1968. 150-154.

15. Meng FR, Bourque PA, Jewett K, Daugharty D, Arp PA. The Nashwaak experimental watershed project: analysing effects of clearcutting on soil temperature, soil moisture, snowpack, snowmelt and streamflow. Water, Air Soil Pollute. 1995; 82: 363-374.

16. Balland V, Pollacco JAP, Arp PA. Modeling soil hydraulic properties for a wide range of soil conditions. Ecol. Model. 2008; 219: 300-316.

17. Päivänen J. The bulk density of peat and its determination. Silva Fennica. 1969; 3: 1-19.

18. Gafni A, Brooks KN. Hydraulic characteristics of four peatlands in minnesota. Can. J. Soil Sci. 1990; 70: 239-253.

19. Silc T, Stanek W. Bulk density estimation of several peats in northern Ontario using the von Post humification scale. Can. J. Soil Sci. 1977; 57: p. 75.

20. Szajdak LW, Szatylowicz J, Kolli R. Peats and peatlands, physical properties. In Gliński K, Horabik J, Lipiec J, editors. Encyclopedia of Agrophysics.: Springer; 2011. 551-555.

21. Letts MG, Roulet NT, Comer NT, Skarupa MR, Verseghy DL. Parametrization of peatland hydraulic propoerties for the Canadian land surface scheme. Atmosphere-Ocean. 2000; 1: 141-160.

22. Dube S, Plamondon A, Rothwell R. Watering up after clear-cutting on forested wetlands of the St. Lawrence lowland. Water Resource Research. 1995; 31: 1741-1750.

23. Boelter DH. Hydraulic Conductivity of Peats. Soil Sci. 1965; 100: 227-231.

24. Boelter DH, Verry E. Peatland and water in the northern lake states. General Technical Report NC-31. St. Paul, MN: Forest Service, North Central Forest Experiment Station, U.S. Dept. of Agriculture; 1977.

25. Lull HW. Soil Compaction on Forest and Range Lands: Forest Services, U.S. Department of Agriculture; 1959.

26. Kramer PJ, Boyer JS. Chapter 4: Soil and Water. In Water relations of plants and soils.: Academic Press, Inc; 1995. 84-114.

27. Kozlowski TT. Soil compaction and growth of woody plants. Scand. J. For. Research. 1999; 6: 596-619.

28. Jutras MF, Nasr M, Castonguay M, Pit C, Pomeroy J, Smith TP, et al. Dissolved organic cardon concentrations and fluxes in forest catchments and streams: DOC-3 model. Ecol. Model. 2011; doi:10.1016/j.ecolmodel.2001.03.035.

29. McNabb DH, Startsev AD, Nguyen H. Soil wetness and traffic level effects on bulk density and air-filled porosity of compacted boreal forest soils. Soil Sci. Soc. J. America. 2001; 65: 1238-1247.

30. Froehlich HA, McNabb DH. Minimizing soil compaction in Pacific Northwest forests. Proceedings Forest Soils and Treatment Impacts Conference. Knoxville, TN; 1984.

31. Jamshidi R, Jaeger D, Raafatnia N, Tabari M. Influence of two ground-based skidding systems on soil compaction under different slope and gradient conditions. Int. J. For. Eng. 2008; 19: 9-16.

Chapter 4

FIVE THINGS YOU DIDN'T WANT TO KNOW ABOUT HYDRAULIC FRACTURES

Vincent M. C.[1]

[1] Fracwell Llc, Golden, Colorado, USA

ABSTRACT

It is common to envision and design hydraulic fractures as if they were simple, planar features that are relatively consistent in width and durable in their flow capacity. Production forecasting is frequently based on a simplified description of the reservoir as a homogeneous single productive layer. In rare instances the pay intervals may be simulated with as many as a dozen layered strata, but even the most meticulous reservoir engineer may mistakenly assign each layer a highly conductive, durable connection with the wellbore. When analyzing the resulting production data, similar assumptions are made, which can erroneously reinforce these misconceptions. Although our industry has been confronted with photographic evidence from minebacks and core-throughs of actual fractures, we have typically failed to incorporate those complexities and challenges into our design, interpretation, and optimization processes. Similarly, we frequently fail to recognize the challenges of highly laminated and highly compartmentalized reservoirs. In many resource plays, hydraulically stimulated horizontal wells appear to be the only completion technique that can achieve economic production rates from these low permeability reservoirs. However the productivity and ultimate recovery from these horizontal wells will be increasingly reliant on durable hydraulic fractures to contact and drain the hydrocarbons through highly laminated formations for the decades necessary to deplete low permeability reservoirs. Oversimplified models typically result in poorly designed completions and missed opportunities. Frequently, the underperformance of a well will be blamed on "poor reservoir quality" instead of correctly recognizing the inadequacy of our created fractures. This paper will examine five limitations of hydraulic fractures and interpretation techniques, and describe the increases in well productivity that can be achieved when efforts are made to address and compensate for these deficiencies.

INTRODUCTION

Early frac engineers certainly recognized that hydraulic fractures were complex features. Geologists, mining engineers, and prison chain gangs all assured us that rocks break in complex manners. But the math is hard, and we aren't capable of predicting nature's complexity. We cannot accurately calculate the pressure losses through a proppant pack with complex geometry, irregular aperture, and with several fluid phases flowing at high velocity. So our predecessors were forced to simplify the description. As a first-order approximation, they assumed that fracs were simple, vertical planes, with uniform width and predictable hydraulic continuity.

Two subsequent generations of petroleum engineers have been introduced to simplified planar hydraulic fractures that have been distorted to fit on a textbook page, such as in Figure 1. Unfortunately, many engineers mistakenly envision fracs as wide, highly conductive channels instead of thin, narrow ribbons of proppant that extend deeply into the reservoir but are vulnerable in their hydraulic continuity. Fractures are commonly modeled to be symmetrical, bi-wing planes that reliably contact the targeted hydrocarbons.

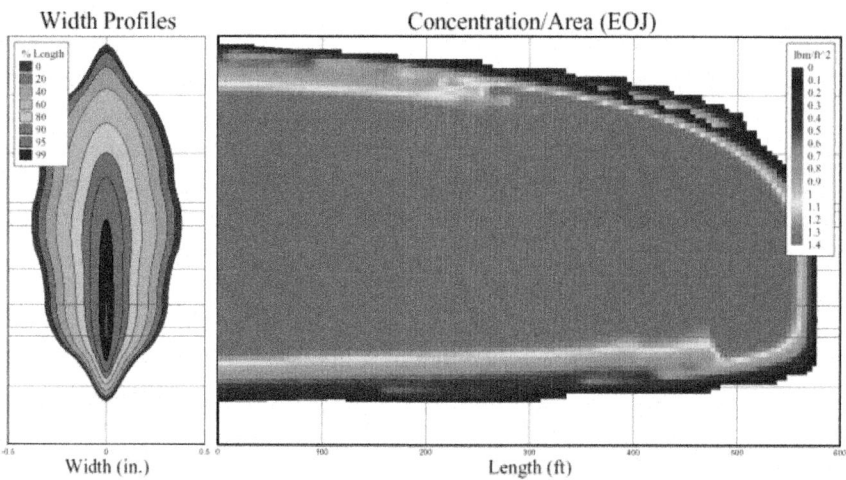

Figure 1. The proportions of fractures are often distorted and misrepresented in simplified models. This figure implicitly assumes the fracture grows symmetrically on either side of the wellbore.

Although chemical engineers clearly recognized that Darcy's flow would not describe pressure losses in porous media, early frac engineers disregarded non-Darcy and multiphase flow effects, and further assumed a single homogeneous reservoir layer was contacted by a highly conductive fracture that permanently connected the wellbore to the hydrocarbons. These assumptions allowed the "optimization" of frac treatments to become a mathematically simple routine. Two subsequent generations of petroleum engineers have filled our literature and conventional wisdom with simulations and "rules of thumb" that would allow us to optimize these mythical ideal fractures. Unfortunately, many of the assumptions are wrong, and our fracs are not optimized.

COMPLEX FLOW REGIMES

Even if fractures were simple, wide features, with perfectly uniform proppant arrangements throughout the entirety of the fracture length and height, our industry would still overestimate the flow capacity of fractures by several orders of magnitude. Figure 2 shows the apparent flow capacity of proppant packs, measured in the laboratory.

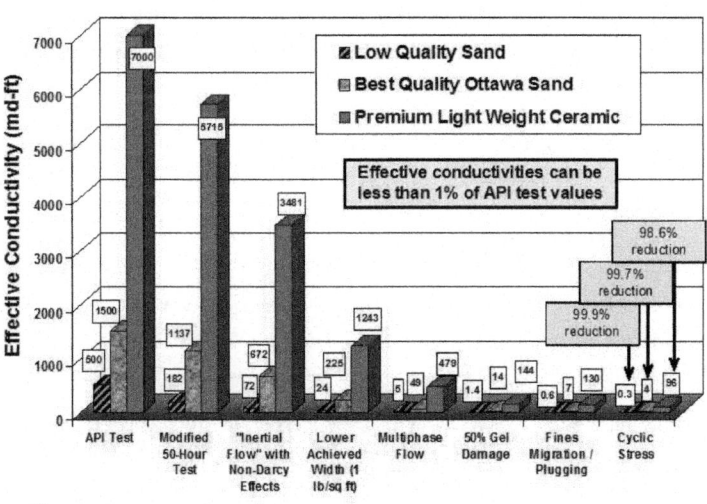

Figure 2. Even in simple, planar proppant packs with uniform proppant distribution, the effective conductivity is frequently 50 to 1000 times lower than published values [1].

Conductivity data provided by most proppant vendors, and utilized in most production simulators are collected with test procedures similar to the left two categories of bar columns in Figure 2. When testing is more sophisticated, with realistic velocities of multiphase fluids through proppant packs subjected to gel damage and cyclic stress oscillations, the pressure losses are often found to be orders of magnitude higher than indicated by reference data [1,2,3].

CONDUCTIVITY DEGRADES

Even the meager amount of effective conductivity shown in Figure 2 appears to be unsustainable. Five different researchers have published the performance of proppants when tested in the laboratory for weeks instead of hours Montgomery [4], McDaniel [5], Cobb [6], Hahn [7], Handren [8]. All five have shown that proppants lose conductivity over time, with one representative test shown in Figure 3. Some proppants are more durable than others, and some laboratory conditions will more rapidly degrade proppant, but not a single proppant pack in the lab has sustained flow capacity without continued particle breakage and compaction during extended testing. The degradation mechanism in these tests has nothing to do with chemical damage, scale deposition, or diagenesis – these conductivity losses are related to the strength of the particles, and show similar trends when tested in dry nitrogen gas, in oil, or in brine, when confined between sandstone, stainless steel, or Teflon. [5, 9, 10]. It is surprising that none of our models incorporate frac degradation over time, despite unanimous evidence that conductivity declines.

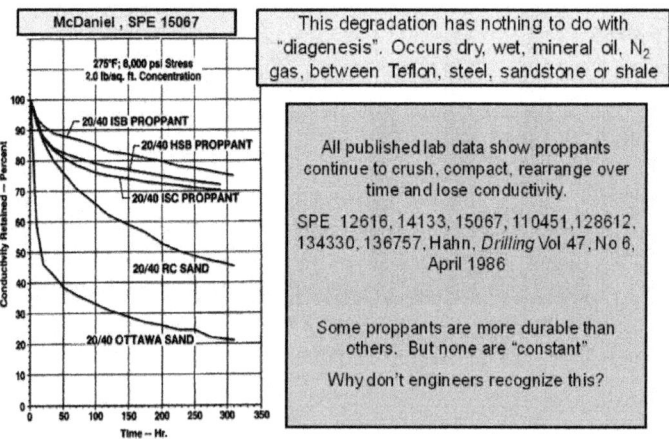

Figure 3. Extended duration tests routinely show continued mechanical crush and loss of flow capacity of proppant packs [5].

HETEROGENEOUS RESERVOIRS

Production forecasting is greatly simplified if the reservoir can be described as a uniform layer with predictable, consistent permeability in the vertical and horizontal directions. However, sedimentary rocks were formed from hundreds or thousands of sequential layers of sediment as shown in Figure 4. Productive lenses can have varying lateral extent.

Figure 4. On every scale, formations may have laminations that hinder vertical permeability and fracture penetration. Shown are thin laminations in the Middle Bakken [11], layering in the Woodford [outcrop photo courtesy of Halliburton], and large scale laminations in the Niobrara [adapted from 12] [13].

The consequences of these laminations are two-fold:

- Vertical perm is terrible. Often the vertical perm is only a tiny fraction of the horizontal perm; k_v/k_h <0.001. Oil and gas do not move easily in the vertical direction through rock. If you want to drain it, you have to frac it. Especially with horizontal wells drilled into a single layer, the frac engineer must create a durable, conductive pathway breaching the laminations within the hydrocarbon-bearing intervals if we have a prayer of draining the reserves from these tight, laminated resource plays, unless pre-existing natural fractures provide a vertical flow path.

- Laminations hinder frac penetration [13]. Fracs don't like to grow through a series of bonded and unbonded layers (Fig 5).

COMPLEX FRAC GEOMETRY

Figure 5 depicts conceptualized fracture branching as it grows through a laminated formation. Figure 6shows minebacks of actual fracturing treatments performed at the Nevada test site and Figure 7 shows a core-through of a treatment in the Piceance Basin of western Colorado.

Figure 5. Instead of perfectly vertical fractures (left) it may be appropriate to anticipate difficulty creating and sustaining a conductive fracture throughout the entire pay
interval [outcrop photo courtesy of Halliburton [13].

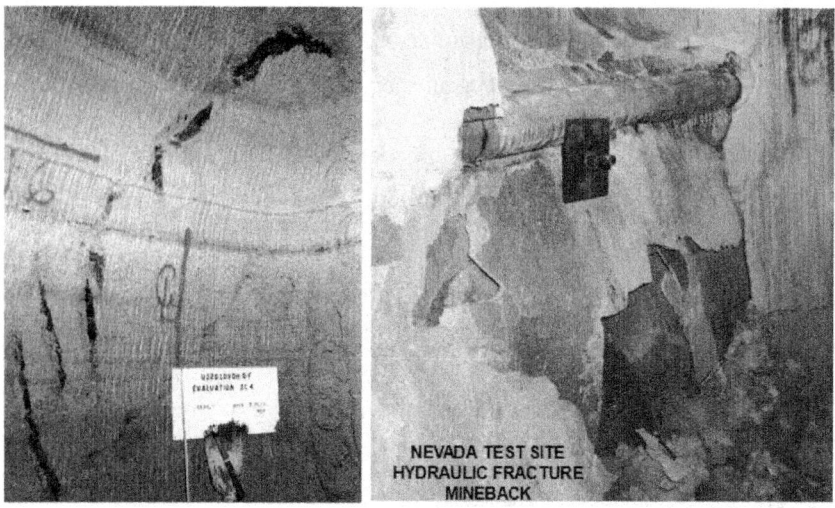

Figure 6. Photographs of mine backs at the Nevada test site demonstrate complexity
[1, 14,15].

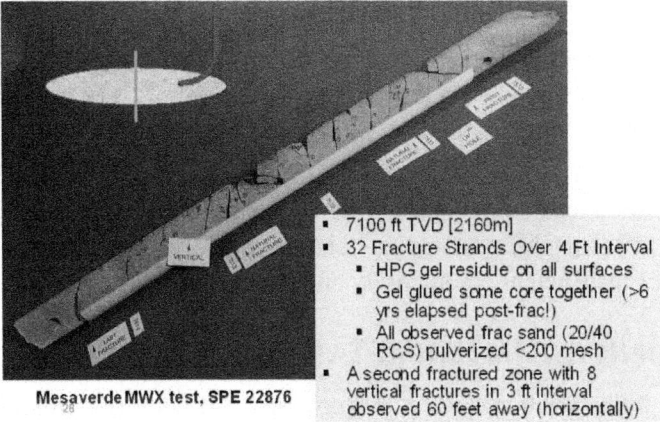

Figure 7. In the Piceance Basin, cores through a created fracture document 40 fracture strands, with only pulverized resin coated sand recovered [1, 16].

Clearly, there is evidence that fractures can grow in much more complicated manners compared to the simple, planar features that are typically presumed in our designs and "optimization" attempts. What are the implications of complexity shown in Figure 8?

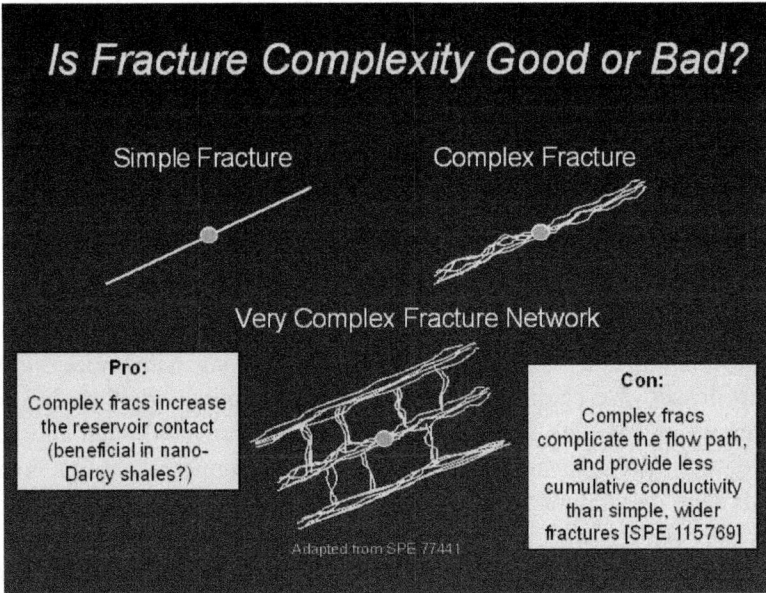

Figure 8. Fracture complexity increases reservoir contact, but challenges our ability to create a durable proppant pack with sufficient hydraulic continuity [adapted from 17].

Hydraulic fractures must achieve two primary objectives. They must:

- Touch rock (contact hydrocarbons)
- Provide a durable conduit for hydrocarbons to flow to the well with acceptable pressure losses (sufficient conductivity)

Complex, branching fractures do an excellent job of touching rock. However, they challenge our ability to place a commensurate degree of conductivity. Branching, complex features are often ineffectively propped, with risk of insufficient conductivity and continuity.

NON-UNIQUE INTERPRETATIONS

The fifth thing we don't want to know about fractures is that it is nearly impossible to identify the deficiencies when analyzing production data from a single well. Figure 9 shows the production history (decline curve and cumulative production) from a single fractured interval, along with three plausible production matches.

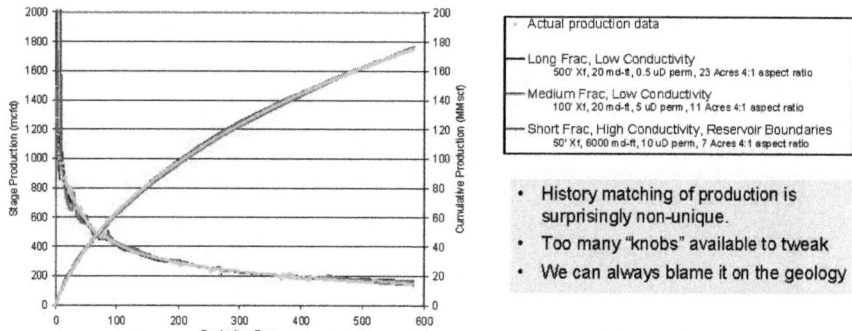

Figure 9. With a single well, the production history can be matched with a nearly infinite combination of plausible fracture and reservoir descriptions [18, 19].

From a single decline curve, we cannot uniquely determine whether the fracture is short and "infinitely conductive," or long with more significant pressure losses. We cannot prove from a decline curve whether the fracture was simple or complex in geometry. We cannot prove whether the fracture conductivity was constant or degrading. Most engineers attempt to match the data with an analytic solution or a numerical simulator that presumes the frac is fully packed with proppant throughout, providing uniform and durable flow capacity without collapse of poorly propped sections. Note that with this approach an engineer can continue to reinforce any existing misconceptions. Fracs can be interpreted to be long or short. Disappointing well productivity can always be blamed on the geology – with no irrefutable proof that the fracture was insufficient.

DISCUSSION OF FIVE DEFICIENCIES

There are certainly more than five deficiencies in our stimulation designs and our techniques to analyze well production. However, the five issues described in this paper include:

- Hydrocarbons move in a complex manner within propped fractures, increasing the pressure losses by 50 to 1000-fold over common expectations, even if the fractures are planar and fully propped.

- Fracture conductivity is not constant. Lab data suggest that all conventional proppant types suffer continued crush and compaction over time.

- Reservoirs are laminated and compartmentalized. Especially with horizontal drilling, ultimate recovery is far more dependent on fracture continuity through laminations than in vertical wells in which each prospective layer can be perforated and individually stimulated. With low perm reservoirs, significantly longer well life (and proppant durability) will be required to drain the available reserves.

- Fractures develop varying degrees of complexity. This is both good and bad. Reservoir contact is increased as fractures branch, twist, and energize pre-existing planes of weakness. However, this complexity challenges our ability to place a durable, hydraulically continuous proppant pack with conductivity commensurate to carry hydrocarbons with an acceptably small pressure loss.

- History-matching of production data is surprisingly non-unique. An engineer can reinforce misconceptions throughout an entire career without encountering any results that cannot be matched with a simple, planar frac of durable, high conductivity in a homogenous reservoir. Underperformance can always be attributed to other factors.

While this is a fairly depressing view of the problem, there are techniques to remove some of the uncertainty and ambiguity allowing significant improvement in the performance of stimulation treatments.

REMOVING THE UNCERTAINTY

Several datasets and techniques can be used to more uniquely describe the performance of propped fractures [19]:

- Wells that are restimulated. When we refrac a well, we have an opportunity to history-match the production from the initial and subsequent stimulation treatments using only a single reservoir description. Difference in well production must be uniquely attributed to the frac design. There have been more than 140 published examples, and history-matching attempts have frequently indicated that fractures are not as effective or durable as previously anticipated [10, 20].

- Fields in which a carefully conducted field trial examines the role of a single variable in fracture design. For instance, when 150 wells are treated at 4 ppg and 150 offset wells are systematically selected to receive 6 ppg slurry, it is possible to achieve comparisons with compelling statistical significance. The difference in productivity is known to relate to the frac performance, and cannot be attributed to reservoir parameters. The evaluation of 200 published field examples [1] provides very credible evidence that fracs do not perform as most people anticipate, and that increased focus on fracture conductivity is merited.

- Wells that are connected by a propped fracture. As described previously, fractures can reach impressive lateral dimensions. It is not uncommon for fractures to intersect adjacent wellbores completed at the exact same depth or in the same formation subinterval. When this occurs, it provides a significant opportunity to investigate the initial and sustained continuity over time. In most cases, adjacent wells appear to lose hydraulic continuity over time, suggesting that the connecting fracture "collapses" or "heals".

- Infill drilling. In many tight reservoirs, we have successfully drilled wells within 200 feet of existing wells and encountered near-virgin reservoir pressures. In many shale reservoirs, infill wells are anticipated to recover nearly 80% of the reserves of adjacent parent wells drilled many years earlier, demonstrating that initial wells have not captured the available reserves.

- More sophisticated modeling and data analyses. While simple production data analyses yield non-unique solutions, several degrees of freedom can be removed with careful analyses of pressure-transient or rate-transient data. There have also been advances in interpretation of flow regimes from wells with complicated fracture networks. Even in the 400-nanoDarcy Barnett shale, production data do not indicate that the entire created network remains highly effective.

These efforts strongly indicate that additional focus on the conductivity, durability and *effectiveness* of the fracture is needed – not just a focus on created dimensions.

OPPORTUNITIES TO IMPROVE FRACTURE PERFORMANCE

It is important to recognize that our intuition, our models, and our traditional interpretations of fracture performance are flawed, and can prevent us from recognizing opportunities to improve well productivity. While our industry has collected data demonstrating complexities (in reservoir description, in fluid flow regimes, in fracture geometry, in durability of proppant packs), the industry has been very slow to adapt designs to accommodate or capitalize on these realities.

More than a dozen specific recommendations have previously been discussed [19] to improve the productivity and profitability of fracturing treatments. However, a general theme is to continue experimenting and studying production from wells, with a healthy skepticism of model predictions and of historic rules of thumb regarding fracture design. Another common finding is that emphasis on improving the effectiveness and durability of treatments appears to be adding more value than blindly focusing on fracture length or treatment volume. There are a great number of field examples in which modest changes to fracturing designs resulted in very large changes to well productivity, convincingly demonstrating that our initial frac designs were insufficient to capture the full well potential. Figure 10 shows surprising increases in productivity were achieved by restimulating a modest perm oil reservoir and a tight gas reservoir with improved fracture designs more focused on the durability and conductivity of the fracturing treatments.

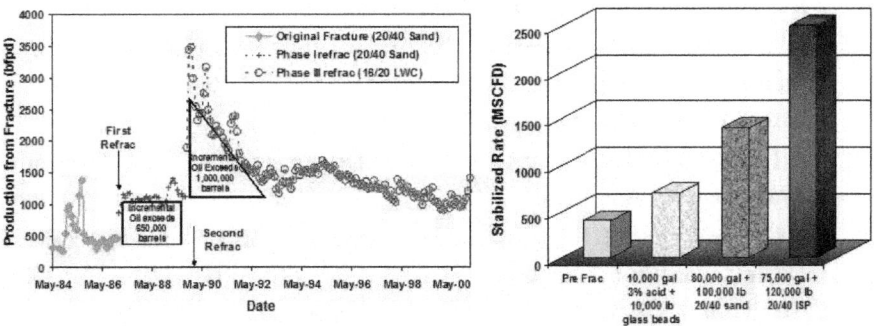

Figure 10. Experimentation with frac design often demonstrates the well potential is constrained by insufficient fracture designs [1, 20].

Similar production increases have been documented in hundreds of field studies in shales, carbonates, coals, and sandstones [1]. On one hand, it is frustrating to admit that after decades we have failed to optimize our fracturing treatments. On the other hand, it is great news that our fracs are not optimized. Reservoirs are often capable of tremendous increases in productivity with improved fracture designs that accommodate and capitalize on our understanding of complexity.

REFERENCES

1.   M. C Vincent, 2009 Examining our Assumptions- Have Oversimplifications Jeopardized Our Ability to Design Optimal Fracture Treatments? Paper SPE 119143 presented at the 2009 Hydraulic Fracturing Technology Conference, The Woodlands, Jan 1921

2.   R. D Barree, S. A Cox, V. L Barree, M. W Conway, 2003 Realistic Assessment of Proppant Pack Conductivity for Material Selection. SPE paper 84306 presented at the Annual Technical Conference, October 58

3.   T Palisch, R Duenckel, L Bazan, H. J Heidt, G Turk, 2007 Determining Realistic Fracture Conductivity and Understanding its Impact on Well Performance- Theory and Field Examples. SPE paper 106301 presented at the 2007 Hydraulic Fracturing Technology Conference, College Station, TX, Jan 2931

4.   C. T Montgomery, and R. E Steanson, 1984 Proppant Selection- The Key to Successful Fracture Stimulation. SPE paper 12616 presented at the Deep Drilling and Production Symposium, Amarillo, TX April 13

5.   B. W Mcdaniel, 1986 Conductivity Testing of Proppants at High Temperature and Stress. SPE Paper 15067 presented at the 56th California Regional Meeting, Oakland, April 24

6.   S. L Cobb, and J. J Farrell, 1986 Evaluation of Long-Term Proppant Stability. SPE paper 14133 presented at the International Meeting on Petroleum Engineering, Beijing, Mar 1720

7.   G Hahn, 1986 How Long will it Prop? Drilling, the Wellsite Publication. 476 Issue 596, April 1986.

8.   P Handren, and T Palisch, 2007 Successful Hybrid Slickwater Fracture Design Evolution. Paper 110451 presented at the 2007 Annual Technical Conference, Anaheim, Nov 1114

9.   R Duenckel, M. W Conway, B Eldred, M. C Vincent, 2011 Proppant Diagenesis-Integrated Analyses Provide New Insights into Origin, Occurrence, and Implications for Proppant Performance. SPE paper 139875 presented at the SPE Hydraulic Fracturing Technology Conference, The Woodlands, TX Jan 2426

10. M. C Vincent, 2010a Refracs- Why, Do They Work, and Why Do They

Fail in 100 Published Field Studies?, SPE 134330 presented at the 2010 Annual Technical Conference, Florence, Italy, Sept 1922

11. LeFeverJ. 2005Overview of Bakken Stratigraphy and "Mini" Core Workshop. AAPG Rocky Mountain Meeting Short Course #1, Jackson, WY, Sept 24.

12. Noble Energy Analyst Conference2010June 3.

13. M. C Vincent, 2011Optimizing Transverse Fractures in Liquid-Rich Formations, SPE 146376 presented at the Annual Technical Conference, Denver, CO Oct 30Nov 2.

14. N. R Warpinski, 1983Investigation of the Accuracy and Reliability of In Situ Stress Measurements Using Hydraulic Fracturing in Perforated, Cased Holes. Proceedings, 24th U.S. Symposium on Rock Mechanics, Texas A&M University, College Station, TX, 773786June 20-22, 1983.

15. N. R Warpinski, L. D Tyler, W. C Vollendorf, and D. A Northrop, Direct Observation of a Sand-Propped Hydraulic Fracture," Sandia National Laboratories Report, SAND810225May 1981

16. N. R Warpinski, et al1993Examination of a Cored Hydraulic Fracture in a Deep Gas Well. SPE 22876, SPEPF Aug 1993150158

17. M. K Fisher, et al2002Integrating Fracture Mapping Technologies to Optimize Stimulations. Paper SPE 77441 presented at the SPE Annual Technical Conference, San Antonio, Sep. 29Oct. 2.

18. M. C Vincent, et al2007Field Trial Design and Analyses of Production Data from a Tight Gas Reservoir: Detailed Production Comparisons form the Pinedale Anticline. SPE paper 106151 presented at the 2007 Hydraulic Fracturing Technology Conference, College Station, TX Jan 2931

19. M. C Vincent, 2012The Next Opportunity to Improve Hydraulic-Fracture Stimulation. JPT Distinguished Author Series, March 2012118SPE 144702.

20. M. C Vincent, 2010Restimulation of Unconventional Reservoirs: When are Refracs Beneficial?. JCPT June 2011, SPE 136757 presented at the Canadian Unconventional Resources & International Petroleum Conference, Calgary, Oct. 1921

Chapter 5

HYDRAULIC AND SLEEVE FRACTURING LABORATORY EXPERIMENTS ON 6 ROCK TYPES

Sebastian Brenne[1], Michael Molenda[1], Ferdinand Stöckhert[1] and Michael Alber[1]

[1] Ruhr-University Bochum, Germany

INTRODUCTION

Hydraulic tensile strength is a crucial value for planning reservoir stimulation and stress measurements. It is used in the classical breakdown pressure (P_b) relation by Hubbert & Willis [1], where P_b is a function of major and minor principal horizontal stresses S_H and S_h, hydraulic tensile strength σ_T and pore pressure P_0:

$$P_b = 3S_h - S_H + \sigma_T - P_0 \tag{1}$$

Options

For hydraulic fracturing laboratory experiments (MiniFrac – MF) under isostatic confining pressure P_m this might be reduced to:

$$P_b = cP_m + \sigma_T - P_0 \tag{2}$$

Options

The coefficient cc should be equal to two when porepressure is neglected. However, many laboratory experiments [2,3] resulted in values of about 1 for c, which might be explained by poroelastic effects.

Thus, when poroelasticity is excluded in the experiments by taking dry samples and sealing off the central borehole by an impermeable membrane (like a polymer tube), one would expect that c equals two and σ_T will be in the range of the tensile strength as determined by other tensile strength tests.

However, experiments with jacketed boreholes (sleeve MiniFrac – SMF) yield remarkable high values for cc (about 6 to 8) and also for σ_T (about 3 to 5 times the tensile strength of the material) [4]. As a consequence we use a linear elastic fracture mechanics approach to evaluate our experiments.

Theory of Hydraulic And Sleeve Fracturing On Hollow Cylinders

Fracture mechanics deal with stress concentrations around fractures and the definition of propagation criteria for fractures. The theory is essentially based on the works of Griffith [5] and Irwin [6], which led to the introduction of the stress intensity factor K.

$$K = \sigma \sqrt{\pi a}$$
(3)

KK represents the magnitude of the elastic stress singularity at the tip of a fracture of the length 2a subjected to a uniform stress σ. With this concept, it is possible to formulate a simple fracture propagation criterion $K = K_c$. The fracture propagates when K reaches a critical value K_c (fracture toughness) with the fracture toughness assumed to be a property of the rock.

Mode I stress intensity factors (K_I) for arbitrary tractions ($\sigma(x)$) applied to the surface of a fracture of the length 2a may be computed by following formula [7,8]:

$$K_I = \frac{1}{\pi\sqrt{a}} \int_{-a}^{a} \sigma(x) \left(\frac{a+x}{a-x}\right)^{\frac{1}{2}} dx$$
(4)

The direction of propagation is the x-axis and the stresses are applied perpendicular to the fracture. As can be seen from equation (4), K_I increases with growing fracture length. A simple, 2-dimensional model was assumed for determination of stress intensity factors at the crack tips of the hydraulically induced fractures in MF and SMF tests.

Two fractures of length a are radially emanating from a circular hole of radius r in an infinite plate subjected to a compressive far field stress of the magnitude P_m. A fluid pressure P_{inj} is acting on the borehole wall and the pressure inside the fractures is either zero (SMF) or equal to the pressure in the borehole (MF: $P_{frac} = P_{inj}$). Stress intensities on the fracture tips can be determined by superposition of stress intensity factors resulting from each loading type [2,3]:

$$K_{I-MF} = K_I(P_m) + K_I(P_{inj}) + K_I(P_{frac})$$
(5)

$$K_{I-SMF} = K_I(P_m) + K_I(P_{inj})$$
(6)

$K_{I-MF/SMF}$ are not only dependent on the fracture length a (cf. Equations (3) and (4)) but also on the borehole radius r

$K_{I\text{-}MF}$ (full pressure in the fracture) gives an upper bound for stress intensities in this geom- etry (actual $K_{I\text{-}MF}$ might be lower due to a negative pressure gradient inside the fracture), while $K_{I\text{-}MF}$ is only induced by the pressure in the borehole and far-field stresses and is therefore substantially lower than $K_{I\text{-}MF}$ (Figure 1).

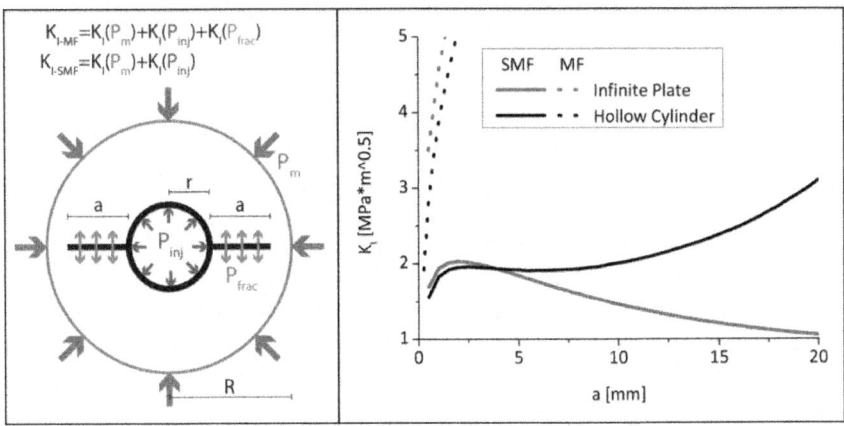

Figure 1. Left side: superposition of stress intensities by each loading type. Right side: stress intensity factor versus fracture length from analytical (infinite plate) and numerical (hollow cylinder) calculations for r =3 mm, P_{inj} =50MPa, P_m =0 and an outer radius R of the hollow cylinder = 30 mm.

As an analytical solution for $K_I(P_m)$ and $K_I(P_{inj})$ for the ring geometry (corresponding to the hollow cylinder) is quite complex, we used the simpler solutions for a circular hole in an infinite plate as described by Rummel and Winter [2,3]. We compared the results of numerical simulations for the ring geometry with analytical solutions for the infinite plate. These results indicate that the simplification might be valid for fracture lengths smaller then $a \approx \frac{R-r}{10}$ with $R=10r$ (R is the outer radius of the ring geometry (cf. Figure 1).

Solving $K_{I\text{-}MF}$ and $K_{I\text{-}SMF}$ for P_{inj} and setting $K_{I\text{-}MF} = K_{I\text{-}SMF} = K_{IC}$ yields a critical injection pressure (P_C (a)) for each crack length a. If P_{inj} reaches P_C (a), the fracture will propagate. From Figure 2 it can be seen, that P_C (a) is very large for very small crack lengths. In consequence, the presence of microcracks is required for the formation of macroscopic fractures.

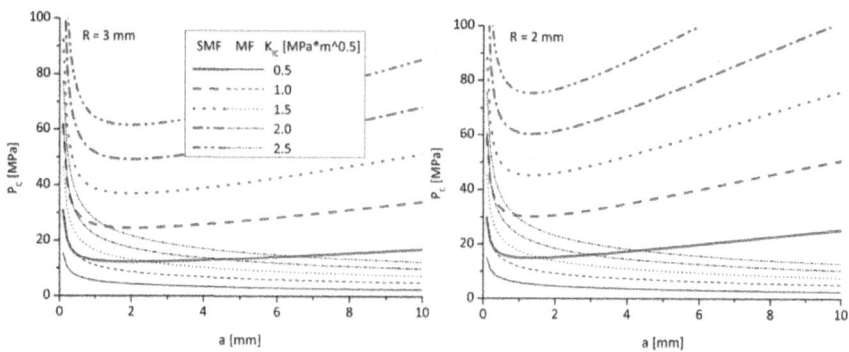

Figure 2. Critical injection pressure for fracture propagation P_C depending on fracture length a for P_m =0. Borehole radii r =3 mm (left), r =2mm (right).

MF-equation (Equation 13) with full injection pressure in the fracture yields unstable fracture propagation at constant injection pressures as soon as microcracks start to propagate. On the other hand, the SMF-equations (Equation 14) show a minimum. Thus, after a fracture reaches the crack length corresponding to the minimum critical injection pressure, stable fracture propagation (i.e. to propagate the fracture, the injection pressure has to be increased) could be expected.

To calculate the coefficient c from Equation 2, we assume the presence of microcracks of a fixed length a_0 in the sample. The corresponding P_C (a_0) versus Pm for the MF case (pressure in fracture = injection pressure) yields a coefficient c =1, which is independent of a_0. P_C (a_0) while for SMF the c value depends strongly on the assumed microcrack length a_0 and gives c >2 (increasing a_0 yield higher c).

SAMPLE PREPARATION AND ROCK TESTING

The core specimens are drilled either with 40 mm or 62 mm water cooled diamond core drills. Core end planes are cut with a water flushed diamond saw blade and ground coplanar to a maximum deviation of ± 0.02 mm. The length and diameter ratio is chosen between 1.5:1 and 2.25:1. After sample preparation core specimens were dried for two days at a temperature of 105°C. For calculations of porosity Φ, measurements of bulk density ρ_d and of grain density ρ_s via pycnometer were done. Static geomechanical parameters were determined by uniaxial and triaxial compressive as well as Brazilian disc tensile strength test series according to ISRM and DGGT suggested methods [9,10].

Mode I fracture toughness was determined using the Chevron notched three-point bending test accord- ing to [9]. Furthermore, a dynamic rock parameter, the compressional ultra-sonic wave velocity (v_p) was measured. For MF/SMF specimens a central axial borehole was drilled into cores, using a water flushed diamond hollow drill with an outer diameter of 4 mm or 6 mm.

Table 1: Rock types used in our experiments.

rock type	era & period	quarry localization	Microstructure
marble	Triassic Upper	Carrara Italy	coarse monocrystalline polygonal fabric
limestone	Jurassic upper Malm	Treuchtlingen South Germany	micritic limestone with abundant fossils and stylolites
sandstone	Carboniferous Mississippian	Dortmund/Hagen West Germany	fine-grained arcose
andesite D	Permian Rotliegend	Doenstedt N German Basin	porphyric fine-grained partly altered and pre-fractured
rhyolite	Permian Rotliegend	Flechtingen N German Basin	porphyric fine-grained partly pre-fractured and sealed joints
andesite R	Permian Rotliegend	Thuringian Forest Rotkopf	porphyric coarse-grained and pre-fractured

Stress Field and Injection

Figure 3 shows schematically the components of the MF and SMF experimental set-up. The stress field is induced by a hydraulic ram (capacity 4500 kN) through a servo controlled MTS Test Star II system with a Hoek triaxial cell which is pressurized using a hand pump to achieve simultaneous pressure increase of confining pressure and axial load. In all tests axial stress is set to be 2.5 MPa higher than P_m to prevent leakage. Distilled water is pumped into borehole as the injection fluid (MF) or into a polymer tube inside the borehole (SMF). A servo controlled pressure intensifier with a maximum injection pressure of 105 MPa was used to perform a constant pumping rate of 0.1 ml/s. With this apparatus also steadystate flow tests were conducted to obtain rock permeability values (according to the procedure described in [11]).

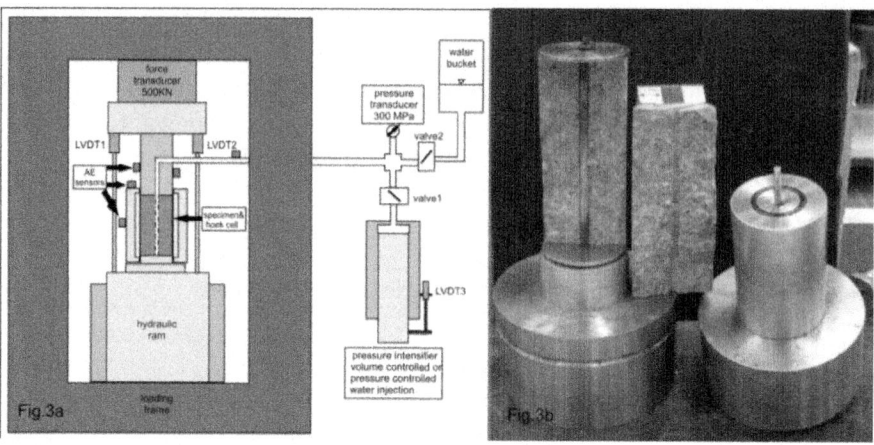

Figure 3: (a) Sketch of MF experimental set up including AE monitoring sensors (not shown are the pressure transduc- er and hand pump system to regulate confining pressure in the Hoek cell). (b) Typical specimen after SMF experiment.

Acoustic Emission Monitoring

Acoustic Emission (AE) signals are acquired with an AMSY5 Acoustic Emission Measurement System (Vallen Systeme GmbH, Germany) equipped with 5 Sensors of type VS150-M. The VS150-M Sensors operate over a frequency range of 100-450 kHz with a resonance frequency at 150 kHz. Due to machine noise in the range below 100 kHz incoming signals are filtered by a digital bandpass-filter that passes a frequency range of 95-850 kHz. AE data are sampled with a sampling rate of 10 MHz. The sensors are fixed using hot-melt adhesive to ensure best coupling characteristics. Pencil-break tests (Hsu-Nielsen source [12]) and sensor pulsing runs (active acoustic emission by one sensor) are used to test the actual sensor coupling on the sample.

RESULTS

Petrophysical and Mechanical Parameters

An overview of the rock properties is given in Table 2. A wide range of low porosity/perme- ability rocks with K_{IC} from 1 to 2 MPam $\sqrt{m}$ were tested.

Table 2. Mean values and standard deviations of petrophysical and mechanical parameters of tested rocks: dry bulk density ρ_d, porosity Φ, permeability k, compressive wave velocity v_p, fracture toughness from Chevron notched threepoint bending tests K_{IC}, cohesion C and friction angle φ from a Mohr-Coulomb fit, Young's modulus Estat , σ_T as determined by Brazilian disc tensile strength tests

rock type	ρ_d [g/cm³]	Φ [-]	k [m²]	v_p [m/s]	KIC [MPa·$\sqrt{m}$]	C/φ [MPa]/[°]	E_{stat} [GPa]	σ_T [MPa]
marble	2.71 ±0.002	0.40 ±0.08	1E-19	5.67 ±0.06	1.57 ±0.11 (N=3)	29/22	36.0 ±1.0	6.4 ± 1.5
limestone	2.56 ±0.008	5.64 ±0.04	1E-18	5.59 ±0.05	1.19 ±0.14 (N=8)	27/53	32.2 ±1.6	8.2 ± 2.2
sandstone	2.57 ±0.006	4.39 ±0.06	8E-18	4.61 ±0.13	1.54 ±0.13 (N=4)	36/50	29.4 ±1.6	13.2 ± 2.1
rhyolite	2.63 ±0.015	1.02 ±0.12	9E-19	5.39 ±0.34	2.16 ±0.10 (N=4)	20...36/55	30.2 ±1.9	15.8 ± 3.2
andesite D	2.72 ±0.023	0.51 ±0.09	6E-19	5.26 ±0.28	1.90 ±0.08 (N=2)	20...41/50	28.7 ±3.1	14.6 ± 4.5
andesite R	2.60 ±0.013	1.70 ±0.08	4E-20	4.35 ±0.27	1.63 ±0.24 (N=4)	31/46	21.3 ±0.9	11.4 ± 2.8

MF and SMF Experiments

A schematic example of typical experiment data for MF and SMF tests is shown in Figure 4. Acoustic emission recordings are used to identify fracture processes in the test speci- mens. AE counts (threshold crossings per time interval – corresponding to AE activity) can directly be linked to localized fracture propagation [4]. The pressure at which the AE count rate raises rapidly is defined as P_{AE}, which is further used as initial fracture propagation pressure. P_{AE} is picked where the AE count rate permanently exceeds 1 / 10 of the test's average (see Figure 4).

In MF experiments, there is almost no AE activity prior to failure. Failure occurs in a very short time span just before sample breakdown (which occurs at maximum injection pressure $P_{inj\ max} = P_b$), therefore in MF experiments $P_{AE} = P_b$. In contrast, SMF experiments show an exponential increase in AE activity at injection pressures that are substantially lower than the actual breakdown pressure ($P_{AE} < P_b$), but much higher than P_{AE} in MF experiments. Therefore, it is possible to interrupt the experiment after AE activity started but before sample breakdown. The latter occurs in SMF experiments when the sample is completely splitted into two parts, which results in a tube breakdown and therefore in an injection pressure drop. Thin sections of specimens, where the experiment was interrupted, show macroscopic fractures emanating several millimeters into the sample but without any connection to the outer surface.

Noteworthy is the discrepancy between the MF and SMF initial fracture propagation pressures P_{AE} at zero confining pressure. This result would imply different hydraulic tensile strength values for the same rock type when using equation (2). Furthermore there is a significant difference between the values of coefficient c calculated for MF and SMF experiments. This

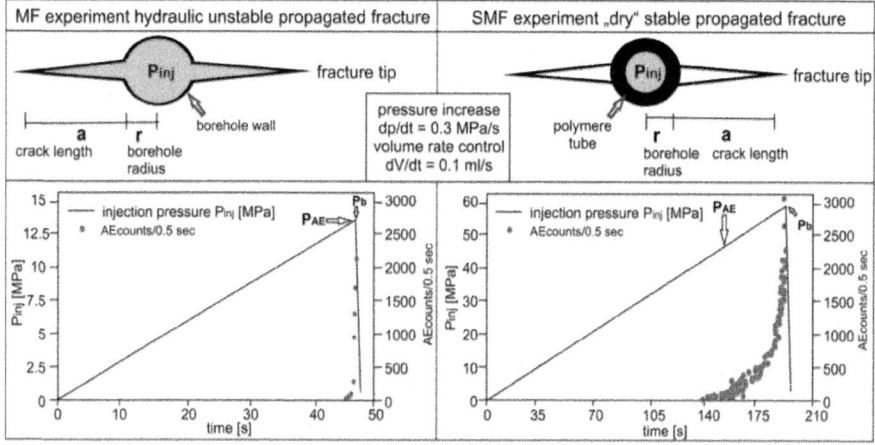

Figure 4: Schematic differences between MF (left) and SMF (right) experiments.

can be seen clearly in Figure 5. Scale effects in P_{AE} (Figure 2) with borehole radius are not evident for the 2 and 3 mm borehole radius samples due to data scattering. One single SMF test of a sandstone with a 6.35 mm borehole radius showed a significantly lower P_{AE} as can be seen in Figure 5.

Table 3: Results of all MF and SMF rock type test series in form of P_{AE0} and coefficient c (see equation (2)). N gives the number of tested samples per lithology and borehole diameter.

rock type	Borehole diam.		MF			SMF	
		N	PAE0 [MPa]	c	N	PAE0 [MPa]	c
marble	4 mm	8	7.7	1.03	6	31.7	6.97
	6 mm	8	9.4	0.96	4	19.6	8.54
limestone	4 mm	9	10.3	1.00	6	26.7	6.06
	6 mm	8	8.2	1.01	7	29.1	5.79
sandstone	4 mm	8	18.2	1.13	5	41.7	6.29
	6 mm	8	18.5	1.14	4	40.5	7.26
rhyolite	4 mm	11	18.2	0.89	4	51.6	6.04
	6 mm	8	16.0	0.85	5	50.9	5.88
andesite D	4 mm	9	16.1	1.00	3	64.2	4.17
	6 mm	6	10.9	0.87	4	48.1	4.83
andesite R	4 mm	10	10.0	1.17	4	47.4	6.26
	6 mm	6	8.2	1.17	5	29.7	7.44
		∑ 93	·	Ø 1.02	∑ 57	·	Ø 6.33

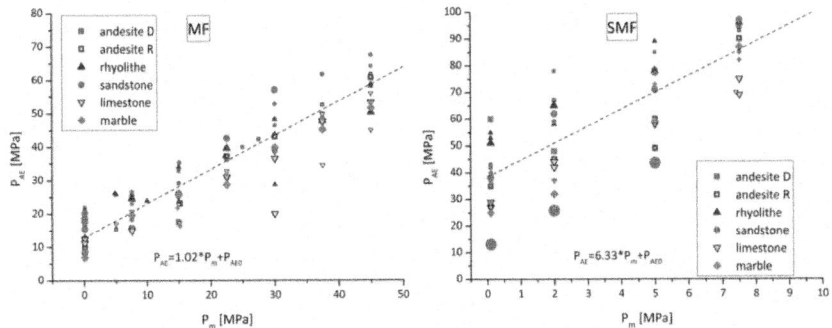

Figure 5: Experimental results of MF (left) and SMF (right) initial fracture propagation pressures for different confining pressures. Dashed line – linear regression of
test data. Symbol size refers to borehole radius r (small – r =2mm; inter- mediate – r
=3mm; large – r =6.35mm)

CONCLUSION

With SMF tests, stable fracture propagation was achieved over a wide range of
injection pressure. Fracture initiation can be confidently linked to the AE count
rates. This can be concluded from experiments that were interrupted after P_{AE}
but below breakdown pressure. Physical examination revealed the presence of
distinct fractures in these speci- mens (see Figure 6).

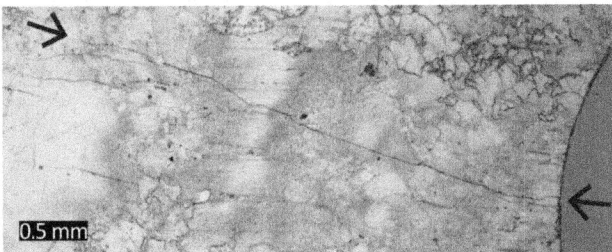

Figure 6: Thin-section of a marble specimen (r =2mm) after SMF test. Clearly visible
is a "dry" fracture (indicated by arrows) emanating radially from the borehole (at the
right side of the picture). The experiment was interrupted before specimen breakdown.
The fracture did apparently not propagate to the outer wall of the specimen.

Due to high data scatter, the theoretical scale effect (critical injection
pressure P_c is higher for smaller borehole radii) cannot be resolved by our data.
However, tests with a larger (r =6.35 mm) borehole give some support to the
notion.

The simple fracture mechanics model is able to explain the higher P_{AE} in SMF experiments. Equations 5 and 6 include the influence of fractures (with or without pressure inside), which is omitted in the classical approach (Equation 1). The high coefficient c in SMF test can only be explained by assuming high microcrack lengths $(a_0 \approx 6 \text{ mm})$.

We excluded poroelastic effects in our analysis due to the use of initially dry rocks with low permeabilities.

Superposition of stress intensity factors for two radial cracks of length a emanating from an internally pressurized (P_{inj}- injection pressure in the borehole, P_{frac}- pressure inside the fracture) circular hole of radius r in an infinite plate subjected to an isostatic far-field stress Pm as described by [2] and [3] :

$$K_I\left(P_m\right) = P_m\sqrt{r} * f_{P_m}\left(a,r\right) \tag{7}$$

$$K_I\left(P_{inj}\right) = P_{inj}\sqrt{r} * f_{P_{inj}}\left(a,r\right) \tag{8}$$

$$K_I\left(P_{frac}\right) = P_{frac}\sqrt{r} * f_{P_{frac}}\left(a,r\right) \tag{9}$$

$$f_{P_m}(a,r) = 2\left(1+\frac{a}{r}\right)^2 \left[\frac{\left(1+\frac{a}{r}\right)^2 - 1}{\left(\pi\left(1+\frac{a}{r}\right)\right)^7}\right]^{\frac{1}{2}} + \left(\pi\left(1+\frac{a}{r}\right)\right)^{\frac{1}{2}}\left(1 - \frac{2}{\pi}\sin^{-1}\left(\frac{1}{1+\frac{a}{r}}\right)\right) \tag{10}$$

$$f_{P_{inj}}(a,r) = \left(1.3\frac{\frac{a}{r}}{1+\left(1+\frac{a}{r}\right)^{\frac{3}{2}}} + \frac{7.8\left(\sin\left(\frac{2a}{r}\right)\right)}{2\left(1+\frac{a}{r}\right)^{\frac{5}{2}} - 1.7}\right) \tag{11}$$

$$f_{P_{frac}}(a,r) = \left(\pi\left(1+\frac{a}{r}\right)\right)^{\frac{1}{2}}\left(1 - \frac{2}{\pi}\sin^{-1}\left(\frac{1}{1+\frac{a}{r}}\right)\right) \tag{12}$$

Note: In equations 10 and 12 the borehole was excluded from the integration of stresses (cf. equation 4). The critical fracture propagation pressure at a given fracture length a, borehole radius r and mode I fracture toughness K_{IC} for the unjacketed $(P_{c\text{-}MF})$ and the jacketed $(P_{c\text{-}SMF})$ case:

ACKNOWLEDGEMENTS

The authors wish to thank the German Federal Ministry for the Environment, Nature Conservation and Nuclear Safety for financing our project (FKZ 0325279B). Many core specimens were prepared and analyzed by our student staff: T. Hoferichter, J. Braun, S. Hönig, K. Bartmann and A. Kraft. A great praise to the precision mechanics workshop guys for the construction of the fine working pressure intensifier system. We appreciate fruitful discussions with geomecon GmbH, Potsdam

REFERENCES

1. Hubbert M, Willis D. Mechanics of hydraulic fracturing. Petroleum Transactions. 1957;210:153–68.

2. Rummel F. Fracture Mechanics Approach to Hydraulic Fracturing Stress Measure- ments. In: Atkinson BK, editor. Fracture mechanics of rock. Academic Press geology series. London

3. .u.a..: Academic Pr; 1987. p. 217–39.

4. Winter R. Bruchmechanische Gesteinsuntersuchungen mit dem Bezug zu hydrauli- schen Frac-Versuchen in Tiefbohrungen. Berichte des Instituts für Geophysik der Ruhr-Universität Bochum: Reihe A. Bochum; 1983.

5. Ito T, Hayashi K. Physical background to the breakdown pressure in hydraulic frac- turing tectonic stress measurements. International Journal of Rock Mechanics and Mining Sciences & Geomechanics Abstracts. 1991;28:285–93.

6. Griffith AA. The Phenomena of Rupture and Flow in Solids. Philosophical Transac- tions of the Royal Society of London. Series A, Containing Papers of a Mathematical or Physical Character. 1921;221:163–98.

7. Irwin GR. Analysis of stresses and strains near the end of a crack traversing a plate. Journal of Applied Mechanics. 1957;24:361–64.

8. Sih GC. Handbook of stress-intensity factors: Stress-intensity factor solutions and formulars for reference. Bethlehem, Pa: Lehigh Univ., Inst. of Fracture and Solid Me- chanics; 1973.

9. Tada H, Paris PC, Irwin GR. The stress analysis of cracks handbook. 3rd ed. New York: ASME Press; 2000.Ulusaihutsen

10. Ulusay R, Hudson JA, editors. The complete ISRM suggested methods for rock char- acterization, testing and monitoring: 1974-2006. 2007th ed. Ankara: Commission on Testing Methods, International Society of Rock Mechanics; 2007.

11. Mutschler T. Neufassung der Empfehlung Nr. 1 des Arbeitskreises "Versuchstechnik Fels" der Deutschen Gesellschaft für Geotechnik e. V.: Einaxiale Druckversuche an zylindrischen Gesteinsprüfkörpern. Bautechnik. 2004;81:825–34.

12. Selvadurai APS, Jenner L. Radial Flow Permeability Testing of an Argillaceous Lime- stone. Ground Water. 2013;51:100–07.

13. ASTM E976. Standard guide for determining the reproducibility of acoustic emsis- sion sensor response. American Society for Testing and Materials. 1994;386-391.

Chapter 6

FRACTURES AND FRACTURING: HYDRAULIC FRACTURING IN JOINTED ROCK

Charles Fairhurst[1, 2]
[1] Senior Consultant, Itasca Consulting Group, Inc, Minneapolis Minnesota, USA
[2] Professor Emeritus, University of Minnesota, Minneapolis, Minnesota, USA

ABSTRACT

Rock in situ is arguably the most complex material encountered in any engineering discipline. Deformed and fractured over many millions of years and different tectonic stress regimes, it contains fractures on a wide variety of length scales from microscopic to tectonic plate boundaries.

Hydraulic fractures, sometimes on the scale of hundreds of meters, may encounter such discontinuities on several scales. Developed initially as a technology to enhance recovery from petroleum reservoirs, hydraulic fracturing is now applied in a variety of subsurface engineering applications. Often carried out at depths of kilometers, the fracturing process cannot be observed directly.

Early analyses of the hydraulic fracturing process assumed that a single fracture developed symmetrically from the packed off-pressurized interval of a borehole in a stressed elastic continuum. It is now recognized that this is often not the case. Pre-existing fractures can and do have a significant influence on fracture development, and on the associated distributions of increased fluid pressure and stresses in the rock.

Given the usual lack of information and/or uncertainties concerning important variables such as the disposition and mechanical properties of pre-existing fracture systems and properties, rock mass permeabilities, in-situ stress state at the depths of interest, fundamental questions as to how a propagating fracture is affected by encounters with pre-existing faults, etc., it is clear that design of hydraulic fracturing treatments is not an exact science.

Fractures in fabricated materials tend to occur on a length of scale that is small; of the order of the 'grain size' of the material. Increase in the size of the structure does not introduce new fracture sets.

Numerical modeling of fracture systems has made significant advances and is being applied to attempt to assess the extent of these uncertainties and how they may affect the outcome of practical fracturing programs. Geophysical observations including both micro-seismic activity and P- and S-wave velocity changes during and after stimulation are valuable tools to assist in verifying model predictions and development of a better overall understanding of the process of hydraulic fracturing on the field scale. Fundamental studies supported by laboratory investigations can also contribute significantly to improved understanding.

Given the widening application of hydraulic fracturing to situations where there is little prior experience (e.g., Enhanced Geothermal Systems (EGS), gas extraction from 'tight shales' by fracturing in essentially horizontal wellbores, etc.) development of a greater understanding of the mechanics of hydraulic fracturing in naturally fractured rock masses should be an industry-wide imperative. HF 2013 International Conference for Effective and Sustainable Hydraulic Fracturing is very timely!

This lecture will describe examples of some current attempts to address these uncertainties and gaps in understanding. And, it is hoped, it will stimulate discussion of how to achieve more effective practical design of hydraulic fracturing treatments.

INTRODUCTION

The term 'rock' covers a wide variety of materials and widely different rheological properties often proximate to each other in the subsurface. Tectonic and gravitational forces, sustained over millions of years, have deformed and fractured the rock on many scales. These forces are transmitted in part through the solid skeleton of the rock, and in part through the fluids under pressure in the pore spaces. Long-term circulation through rock at high temperatures at depth involves dissolution and precipitation along the fluid pathways, producing changes in the chemical composition of the fluids and modifying the overall fluid circulation.

Rock in situ is 'pre-loaded' and in a state of changing equilibrium. Any engineering activity changes this equilibrium (see Appendix 1). Often the changes can be accommodated in stable fashion, but serious instabilities can develop.

The rock mass is opaque. Although geophysics is making impressive advances in defining large structures such as faults and bedding planes, most of the features that influence the rock response to engineering activities remain hidden. Mining and civil engineering activities allow three-dimensional access to the underground and direct observation of smaller features such as fracture networks, but most of the newer engineering applications involve essentially one-dimensional access by borehole. Rock engineering problems fall into the 'data –limited' category, as defined by Starfield and Cundall (1988), and strategies to address them must follow a different strategy than engineering problems where detailed and precise design information is available.

Faced with such complexity and lack of structural details, traditional subsurface engineering design has been guided by empirical procedures developed and refined through long experience.

Projects are now venturing well beyond current experience, and for many, 'novel' applications now considered (e.g., Enhanced Geothermal Systems, Carbon Sequestration, see Appendix 1). There is little experience, few guiding rules and very little data to guide the engineering approach.

Such obstacles notwithstanding, subsurface processes, both long–term geological and short term responses, to engineering activities do obey the laws of Newtonian Mechanics.

Classical continuum mechanics has long been used to guide some aspects of design, but considerable care is required in practical application, due to the need to simplify the representation of the real conditions in order to obtain analytical solutions.

The remarkable developments in high-speed computation and associated modeling techniques over the past one to two decades provide an important new tool, which complemented by the appropriate field instrumentation, can augment the classical continuum analyses and help overcome the lack of prior experience. Some empiricism and general practical guidelines may still be useful for the design engineer, but these can and should be mechanics-informed.

This lecture attempts to illustrate the 'mechanics-informed' approach with respect to the practical application of hydraulic fracturing and related engineering procedures to rock engineering.

HYDRAULIC FRACTURING

Hydraulic fracturing first was used successfully in the late 1940's to increase production from petroleum reservoirs (Howard and Fast, 1970). The technology has evolved since and is now a major, essential technique in oil and gas production. This and other impressive oil industry developments, such as directional drilling, have attracted interest in application of these technologies to a variety of other subsurface engineering operations. Enhanced Geothermal Energy (EGS) is a notable example. Geothermal Energy is a huge resource. Commenting on the EGS resource in the USA, Tester et al. (2005), state:

"….we have estimated the total EGS resource base to be more than 13 million exajoules (EJ)[1] - . Using reasonable assumptions regarding how heat would be mined from stimulated EGS reservoirs, we also estimated the extractable portion to exceed 200,000 EJ or about 2,000 times the annual consumption of primary energy in the United States in 2005. With technology improvements, the economically extractable amount of useful energy could increase by a factor of 10 or more, thus making EGS sustainable for centuries." [2] -

"At this point, the main constraint is creating sufficient connectivity within the injection and production well system in the stimulated region of the EGS reservoir to allow for high per-well production rates without reducing reservoir life by rapid cooling." [3] -

Field experiments to extract geothermal energy from rock at depth by hydraulic fracturing were started in 1970 by scientists of the Los Alamos National Laboratory, USA. Two boreholes were drilled into crystalline rock (one 2.8 km deep, rock temperature 195°C; the other 3.5 km rock, 235°C) at Fenton Hill, New Mexico. Hydraulic fracturing was used to develop fractures from the boreholes in order to create a fractured region through which water could be circulated to extract heat from the rock. The experiment was terminated in 1992. Commenting on what was learned from the Fenton Hill study, Duchane and Brown (2002) note:

"The idea that hydraulic pressure causes competent rock to rupture and create a disc-shaped fracture was refuted by the seismic evidence. Instead, it came to be understood that hydraulic stimulation leads to the opening of existing natural joints that have been sealed by secondary mineralization. Over the years additional evidence has been generated to show that the joints oriented roughly orthogonal to the direction of the least principal stress open first, but that as the hydraulic pressure is increased, additional joints open."

This is an early indication that pre-existing fractures mass significantly affect how hydraulic fractures propagate in a rock mass.

INFLUENCE OF FRACTURES AND DISCONTINUITIES ON THE STRENGTH OF BRITTLE MATERIALS

Hydraulic fracturing can be considered as a technique to overcome the strength of a rock mass in situ, initiation and propagation of a crack through a system of pre-existing fractures, essentially planar discontinuities (e.g., bedding planes), and intact rock.

In examining the fracture propagation process, the pioneering work of Griffith (1921, 1924) is a logical point of departure. Griffith had identified planar discontinuities, or flaws, in fabricated materials as the reason why the observed technical strength of brittle materials was about three orders of magnitude lower than the theoretical inter-atomic cohesive (tensile) strength. [4] - Using an analytical solution by Inglis (1913) for the elastic stresses generated around an elliptical crack in a plate, Griffith observed that the maximum tensile stress at the tip of the crack $\sigma_t = \sigma_0 (1 + 2a/b)$, where a and b are the major and minor semi-axes of the ellipse, and as the ellipse degenerated to a sharp crack or flaw (i.e., as the ratio a/b became very high)[5] - , the stress σ_t could rise to a value high enough to reach the inter-atomic cohesive strength sufficient to cause the original crack to start to extend.

But would the crack continue to extend and lead to macroscopic failure? To address this question, Griffith invoked the *Theorem of Minimum Potential Energy*, which may be stated as "The stable equilibrium state of a system is that for which the potential energy of the system is a minimum." For the particular application of this theorem to brittle rupture, Griffith added the statement, "The equilibrium position, if equilibrium is possible, must be one in which rupture of the solid has occurred, if the system can pass from the unbroken to the broken condition by a process involving a continuous decrease of potential energy."[6] -

Griffith's classical work has provided the foundation for the field of "Fracture Mechanics" [Knott (1973); Anderson (2005)] responsible for major continuing advances in the development of high-performance fabricated materials.

Since we will make reference later to this specific definition by Griffith, it is useful to re-state it here.

THEOREM OF MINIMUM POTENTIAL ENERGY

"The stable equilibrium state of a system is that for which the potential energy of the system is a minimum. The equilibrium position, if equilibrium is possible, must be one in which rupture of the solid has occurred, if the system can pass from the unbroken to the broken condition by a process involving a continuous decrease of potential energy."

Although much of classical Fracture Mechanics has emphasized applications to problems of Linearly Elastic Fracture Mechanics (LEFM) it is important to recognize that the theorem of minimum potential applies equally to inelastic problems.

MECHANICS OF HYDRAULIC FRACTURING

As used classically in petroleum engineering, hydraulic fracturing involves sealing off an interval of a borehole at depth in an oil or gas bearing horizon, subjecting the interval to increasing fluid pressure until a fracture is generated, injecting some form of granular proppant into the fracture as it extends a considerable distance from the borehole into the petroleum bearing formation, and then releasing the pressure. This causes the sides of the fracture to compress onto the proppant, creating a high-permeability pathway to allow oil and/or natural gas to flow back to the well and to the surface.

Figure 1 shows a simple two-dimensional cross-section through an idealized hydraulic fracture. The borehole injection point is at the center of the fracture, which is assumed to be a narrow ellipse that has extended in a plane normal to the direction of the maximum[7] - (least compressive) in-situ stress.

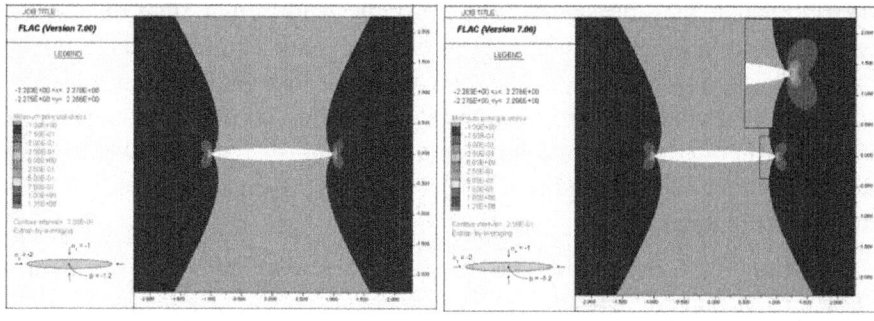

Figure 1: Left) Major and (right) minor principal stresses in the vicinity of an internally pressurized elliptical crack in an impermeable rock.

In the case shown, the crack major/minor axis ratio a/b is 10:1. The internal fluid pressure p = 1.2, while the least compressive principal stress σ_x = 1.0. This results in a tensile stress concentration at the crack tip. The magnitude of the elastic stress concentration at the crack tip increases directly with 2a/b, (Inglis, 1913). Hence for the case of a>>b, i.e., a 'sharp' crack[8] - , the concentration is very high, and the crack will extend essentially as soon as the fluid pressure exceeds the magnitude of the least compressive principal stress (σ_x in Figure

3) it begins to extend, and there will be a pressure gradient from the injection point towards the crack tip as the fluid flows towards the tips. This gradient will depend on the fluid viscosity. Also, since the rock will exhibit some level of permeability, fluid will also flow (or 'leak–off') into the formation as it flows under pressure along the fracture; the rock has a finite strength, or 'toughness' so that energy will be required to extend the crack.

An analytical solution for the stresses in the elastic medium and the crack-opening displacement along the crack was first published by Inglis (1913) and served as the basis for early applications to hydraulic fracturing and fracture treatment design. The Perkins, Kern (1961) and Nordgren (1972) (PKN) andGeertsma and de Klerk (1969) (GDK) models are still used, although numerical models and combinations are now popular. Details of the PKN and GDK models can be found on the SPE website: http://petrowiki.spe.org/ Fracture_propagation_models. Several differences between the stationary crack assumed by Inglis (1913) and a hydraulic fracture introduce significant difficulties in developing an accurate model of the fracturing process. Thus, the fracture is generated by application of an increasing fluid pressure until the fracture is initiated and extends away from the injection point. Flow of fluid in the fracture is governed by classical fluid flow equations of Poiseuille and Reynolds (lubrication); the pressure drop along the fracture depends on the viscosity of the fluid, and the permeability of the rock (leading to fluid 'leak-off'); the fracture aperture depends on the stiffness of the rock mass and the fluid pressure distribution along the crack; and fracture extension depends on the mechanical energy supplied to the region around the crack tip. The tip may propagate ahead of the fluid, leading to a 'lag,'a dry region between the crack tip and fluid front.

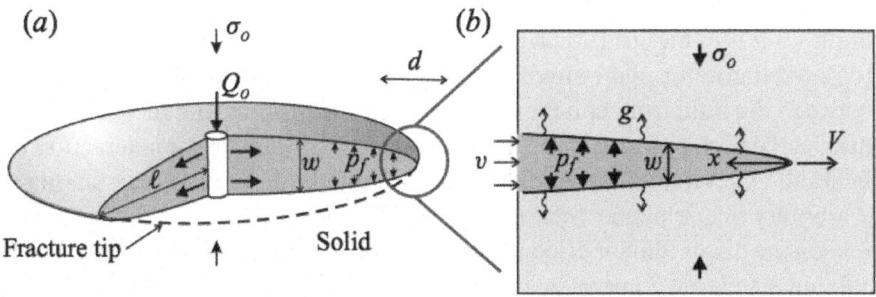

Figure 2: Radial Model of Axi-symmetric Flow and Deformation associated with Hydraulic Fracturing.

Figure 2 illustrates these features for the classical Radial Model in which it is assumed that the fracture propagates symmetrically away from the borehole in a plane normal to the minimum (least compressive) principal in-situ stress, σ0.

Development of efficient and robust Hydraulic Fracturing (HF) simulators is central to successful practical HF treatment of petroleum reservoirs. As noted earlier, competing physical processes are operative during the fracturing operation. This has led to a sustained effort over many years to understand and map the multi-scale nature of the tip asymptotics that arise as a result of these competing physical processes in fluid-driven fracture. These asymptotics solutions are critical to the construction of efficient and robust HF simulators. For example, in an impermeable medium, the viscous energy dissipation associated with driving fluid through the fracture competes with the energy required to break the solid material. Breaking of the bonds corresponds to the familiar asymptotic form of linear elastic fracture mechanics (LEFM), i.e., the opening in the tip region is of the form, e.g., (Rice, 1968), with denoting the distance from the tip. However, under conditions where viscous dissipation dominates, the coupling between the fluid flow and solid deformation leads to (Spence and Sharp, 1985; Lister, 1990; Desroches et al., 1994), on a scale that is considerably larger than the size of the LEFM-dominated region, but still small relative to the overall fracture size. In other words, in the viscosity-dominated regime, the zone governed by the LEFM asymptote is negligibly small compared to the crack length. Thus, in the viscosity-dominated regime, the HF simulator should embed a 2/3 power law asymptote rather than the classic 1/2 asymptote of LEFM. Garagash et al.(2011) discuss the generalized asymptotics near the tip an advancing hydraulic fracture, an extension of two particular asymptotics obtained at Schlumberger Cambridge Research Laboratory in the early 1990's (Desroches et al., 1994; Lenoach, 1995).

Three classes of numerical algorithms for HF simulators have now been built: (i) a moving grid for KGD, radial, PKN and P3D fracture simulators; (ii) a fixed grid for plane strain and axisymmetric HF with allowance for a lag between the fluid front and the crack tip, and fracture curving (a versatile code has been developed at CSIRO[9] - Melbourne to simulate the interaction of a hydraulic fracture with other discontinuities); and (iii) fixed grid for simulating a arbitrary shape planar fracture in a homogenous elastic rock. These codes rely on the displacement discontinuity method (Crouch and Starfield, 1983) for solving the elastic component of the problem, i.e., the relationship between the fracture aperture and the fluid pressure.

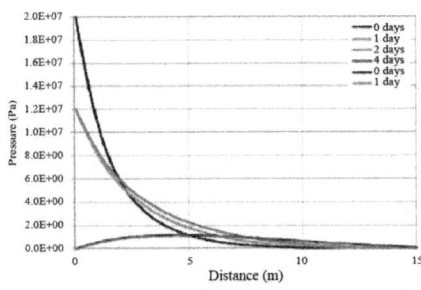
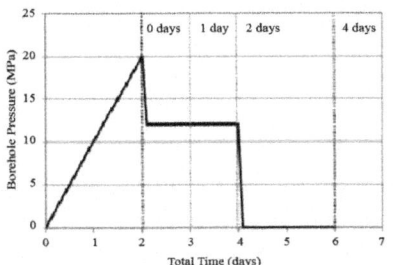

Figure 3: Fluid Pressure Distribution along the Central Axis (Ox) of Figure 1 for a permeable rock due to pressurization and de-pressurization of the borehole.

Figure 3 is presented to illustrate that the fluid pressure in a permeable rock can continue to flow away from the point of injection even after the borehole pressure is reduced to zero. The example shows the distribution of fluid pressure in the rock mass (permeability 5 mD) after (i) 2 days of pressurization up to the peak pressure of 20 MPa in the fracture; (ii) stop pumping and reduce fluid pressure quickly to 12MPa at the point of injection; (iii) hold the pressure constant for 2 days; and (iv) drop the pressure to zero.

It is seen that the pressure in the rock (red curve) has a maximum at some distance from the borehole such that fluid continues to flow into the rock for some time after the pressure in the borehole is reduced to zero. Different combinations of rock permeability, pumping rates and durations can lead to higher peak pressure values in the rock, and longer periods during which fluid can continue to flow away from the well. Such flow may contribute to slip on pre-existing fractures after the pressure in the borehole is reduced to zero.

HYDROSHEAR

Hydraulic fracturing is considered to be initiated from a packed–off interval borehole when the net state of stress around the well bore reaches the tensile strength of the rock. It is important to recognize that fluid pressurization of a well in permeable rock will result in flow of the fluid into the rock as soon as the fluid pressure stimulation process is started. This changes the effective stress state in the rock mass and can lead to slip on pre-existing fractures at fluid pressures below the pressure required to crate and extend a hydraulic fracture. This process of inducing slip on pre-existing fractures is termed 'Hydro-shear'. Flow of pressurized fluid into the rock reduces the effective normal stress $(\sigma n - p)$ everywhere in the rock { σn = normal stress at any point; p = fluid pressure.] If c and μ respectively represent the cohesion and coefficient of friction acting across the surfaces of a fracture in the rock, then the effective resistance of the fracture to (shear) sliding, τr, will be:

$$\tau r = c + \mu\,(\sigma n - p) \tag{1}$$

Thus, if the pressure p is raised progressively then τr will be reduced correspondingly until it reaches the limit at which sliding will occur. The situation is illustrated graphically in Figure 3. The rock is subjected to a three-dimensional state of stress represented by the principal stresses σ1, σ2, σ3 and the fluid pressure p. The series of points 'X' indicate the effective state of stress on an array of pre-existing fractures in the rock. As illustrated in Figure 5, the effect of increasing the fluid pressure in the medium is to move the stress state on these cracks close to the limiting shear resistance, i.e., to the limiting value represented by the Mohr-Coulomb limit. As the stress state reaches this limit, the cracks will slip.

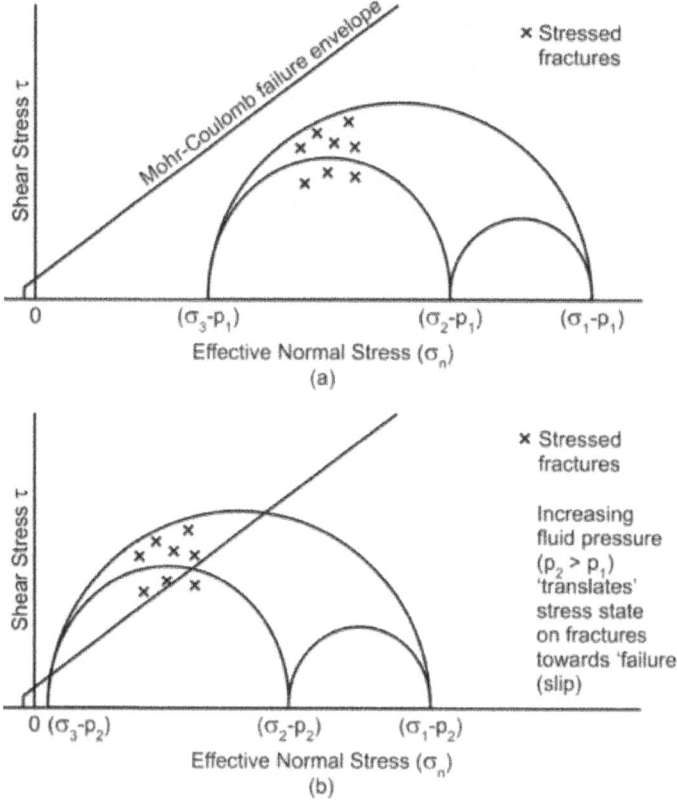

Figure 4: Hydro-shearing — a procedure to generate slip on pre-existing fractures by increasing the fluid pressure to a level below that required to generate a hydraulic fracture.

In order to initiate a hydraulic fracture, the fluid pressure would need to be increased further, until the limiting Mohr circle reaches the tensile strength limit of the failure envelope. Since crack surfaces are often not smooth, shear slip will tend to result in crack dilation, and an associated increase in fluid conductivity. It is suggested that hydro-shearing could be more effective than hydraulic fracturing as a stimulation technique in certain applications, e.g., in stimulation of high-temperature geothermal reservoirs. Cladouhos et al. (2011) discuss the application of hydro-shearing as a geothermal stimulation technique. The possibility that silica proppant may dissolve in the aggressive high-temperature fluid environment of some geothermal reservoirs whereas slip on rough fractures develops aperture increase without the need for proppant is also presented as an argument in favor of hydroshearing.

DEFORMATION AND FAILURE OF ROCK IN SITU

As with fabricated materials, the deformation and failure of brittle rock is also dependent strongly on fractures and discontinuities. In a rock mass, however, the fractures occur over a very wide range of scales from sub-microscopic to the size of tectonic plates. A large specimen of rock will probably include some large fractures, and as the scale of the rock mass increases, fractures from different tectonic epochs.

Study of fracture systems underground in mines and in civil engineering projects allow systems of fractures to be identified and classified statistically into discrete fracture networks (DFN's). The network will include intersecting sets of planar fractures, but individual fractures will tend to be of different lengths, and though organized in two or three spatial orientations, of variable, finite length and not collinear.

Figure 7 presents a two-dimensional illustration of the application of DFN's to the numerical modeling of a fractured rock mass. The in-situ rock mass is considered as a large specimen of intact rock that has been transected by the DFN determined from field observations and fracture mapping underground or at surface outcrops. The properties of the intact rock are built into a Bonded Particle Model of the rock (using the Particle Flow Code (*PFC*) code) based on results of laboratory tests of the intact rock deformability and strength. The intact rock representation is shown on the left of Figure 6. The DFN (shown on the upper right in Figure 6) then is superimposed onto the intact rock.

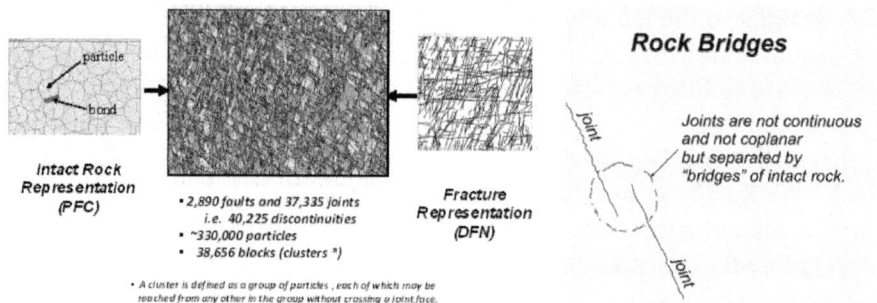

Figure 5: The Synthetic Rock Mass (SRM) representation of a fractured rock mass (in two dimensions).Damjanac et al. (2013) present a discussion of the 'construction' of an SRM in three dimensions.Pierce (2011) presents a comprehensive discussion of practical guidelines and factors involved in the construction of DFN's.

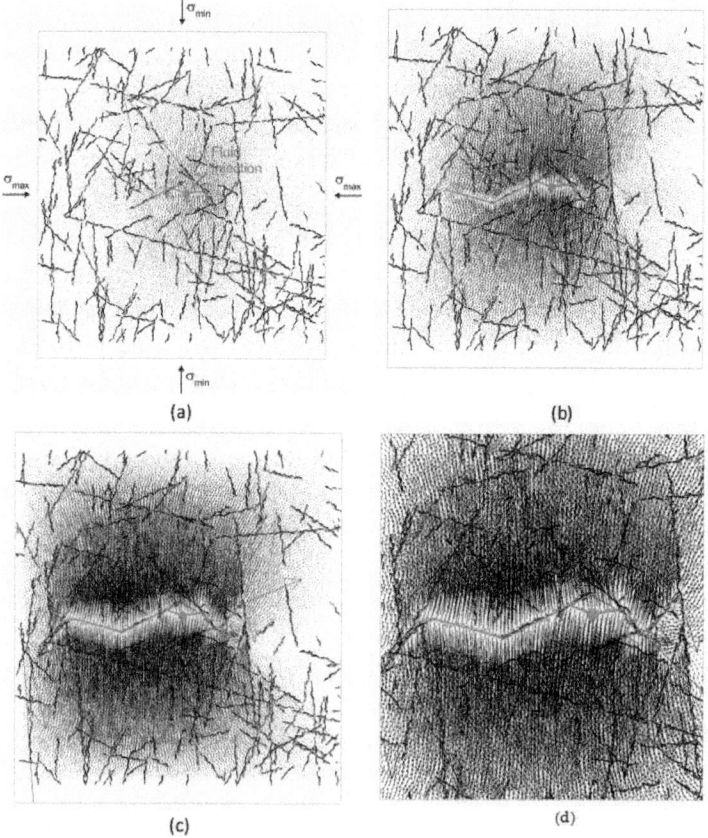

Figure 6: Extracts from simulation of the propagation of a hydraulic fracture in a two-dimensional impermeable SRM (Synthetic Rock Mass). (The horizontal stress σ_{max} is

29 MPa and the vertical stress σ_{min} is 12 MPa – Figure 5(a)). Note that the intact rock between the fractures has a finite strength and can break by rupture of the cemented bonded particles shown in Figure 5. The pressure required to propagate the fracture after breakdown was approximately 10 MPa above the minimum (i.e., least compressive) principal.

Cohesion and friction values are assigned to the joint planes.[10] - The 'unconfined' strength of a typical large SRM is of the order of a few percent of an intact rock specimen of the same rock (Cundall, 2008). Much of the in-situ strength is derived, of course, from the in-situ stresses imposed on the SRM in situ. One of the consequences of the finite length and lack of collinearity of joint sets in DFN's is the formation of bridges of intact rock Figure 4 within the SRM. These bridges provide regions of intact rock, and of stress concentration, in the SRM and account for a significant part of the overall strength of the rock mass. Earlier models of a rock mass, considered to consist of several sets of through-going fractures, exhibited much lower rock mass strength (Hoek and Brown, 1980).

Figure 5 presents selected extracts from a two–dimensional *PFC* simulation of the development of a hydraulic fracture in a jointed Synthetic Rock Mass. The SRM model was developed following the procedure outlined in Figure 5. The joint distribution was based on a DFN obtained at the Northparkes Mine in Australia.[11] - Figure 5(a) shows the location of a vertical borehole that was pressurized by fluid until a hydraulic fracture was initiated. The rock mass is assumed to be impermeable. (The path of the fracture has been traced in blue for clarity.) Displacements in the rock mass produced by the hydraulic fracture are shown as vectors on each side of the fracture. It is seen that the fracture started more or less symmetrically on each side of the borehole, but propagation of the right wing was arrested when the hydraulic fracture encountered an adversely oriented pre-existing joint (Figure 5(b)). With increasing pressure, in the borehole, the hydraulic fracture continued to extend asymmetrically towards the left (Figures 5(c) and 5(d) Figure 5(d) is simply an enlarged view of Figure 5(c)). It is seen that the propagating fracture extended partially by opening existing fractures and partially by developing new fractures through intact rock. Although local deviations occur, the overall path of fracture growth is approximately perpendicular to the direction of the minimum compression stress. The existing fractures introduce an asymmetry to the rock mass. In terms of the idealized symmetric crack of Figure 2, the system in Figure 3 can be considered as two cracks, one extending to the right and one to the left of the borehole with a higher 'fracture toughness' on the right compared to the left, etc.

Jeffrey et al. (2009) conducted an underground test in the Northparkes Mine, Australia to observe the propagation of a hydraulic fracture in naturally fractured tock. Figure 7 shows part of the path of the fracture, as seen in a tunnel excavated into the fractured rock. The fracture path shows similar characteristics to those shown in the *PFC* simulation in Figure 6.

Figure 7: Hydraulic fracture (green plastic) crossing a shear zone on the face of a tunnel excavated through the fracture. "The arrows indicate the trace of the fracture with green plastic contained in it. There is no clear fracture between points 1 and 2 but the fracture may have crossed this zone either deeper into the rock or in the rock that has been excavated. Approximately 2 m of fracture extent is visible" (Jeffrey et al., 2009).

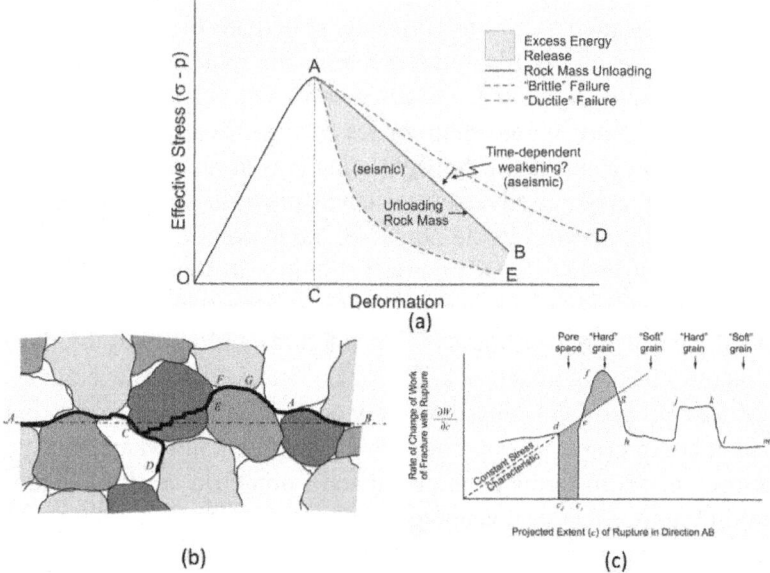

Figure 8: Energy changes during propagation of a fracture through heterogeneous rock.

The energy required to initiate crack propagation is represented by the area OAC in Figure 7(a). Whether or not the crack will extend depends on the energy that becomes available from the intact rock around the crack. If the energy released from the rock mass, represented by the area under the red curve AB, is greater than the energy required to extend the crack, represented by the area under curve AE, then the crack will extend; the excess energy represented by the shaded area serves to accelerate the crack and release seismic energy. If the energy required to extend the crack is represented by the area under the green curve AD, it is greater than the energy that would be released from the rock mass, and hence the crack would not extend. It is possible that the crack could exhibit some form of time-dependent weakening (e.g., due to fluid flow to the crack, viscous behavior, etc.) such that the energy required to extend the crack would be reduced. This could lead to crack extension, i.e., as the slope AD increased to overlap AB, but with no excess energy to produce seismicity. Figures 7(b) and 7(c)[12] - illustrate another feature of crack extension on the granular scale. The energy required to extend a crack through or around a grain will be variable; the fracture may encounter pore spaces where no crack energy is required. Application of a constant load to such a heterogeneous system will result in local acceleration and deceleration of the crack-producing bursts of microseismicity. Similar effects can arise in rock fracture propagation at all scales.

It is worth noting that all of these processes of fracture propagation, albeit complex, develop in accordance with the principle of seeking the minimum potential energy of the system.

Much of the preceding discussion has focused on two-dimensional analysis or models. In reality, we are dealing with three- dimensional space (as noted in Figure 6), plus the influence of time (e.g., with respect to fluid flow, or time-dependent rock properties). Figure 8 provides an example from an actual record of hydraulic fracture propagation.

Figure 8 shows the sequence of microseismic events observed during hydraulic fracture stimulation ('treatment' in Figure 8(a)) of a borehole. Early time events are shown as green dots; later events are in red. The microseismic pattern indicates that fracturing started on both sides of the borehole at the injection horizon, but then moved up some 100 m to a higher horizon. As pumping continued, fracturing continued (red locations) on both horizons. It was concluded that the initial fracture in the lower horizon had intercepted a high-angle fault, allowing injection fluid to move to the higher level where it opened up and extended another fracture. Continued pumping led to fracture extension on both horizons. Numerical analysis Figure 8(b) indicated that initial fracture propagation at the lower level resulted in induced tension on the fault above the horizon, but compression on the fault below the lower injection horizon. This explains why injection fluid did not penetrate along the fault below the horizon, and provides a good illustration of the benefit of combining numerical analysis with field observation in understanding fracturing processes.

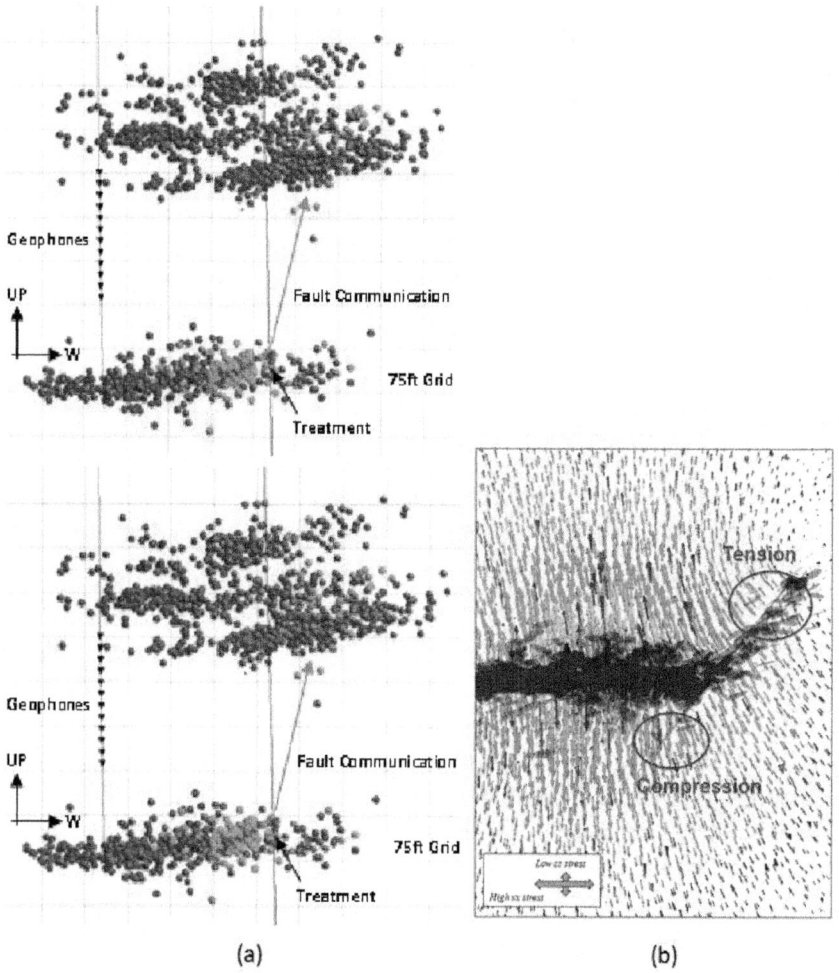

(a)   (b)

Figure 9: a) Microseismicity observed during hydraulic fracturing in a deep borehole; (b) numerical 'explanation' of the behavior observed in (a).

MICROSEISMICITY AS AN INDICATOR OF SLIP ON FRACTURES

Microseismicity stimulated during hydraulic fracturing and associated stimulation techniques (e.g., hydroshear) is often used to indicate slip and deformation on fractures in the rock. In some cases, it is tacitly assumed that absence of microseismicity indicates absence of slip or deformation. In fact, there is growing evidence that microseismicity does not present a complete picture of deformations induced by stimulation or other effects leading to stress change. Figure 9, reproduced from Cornet (2012) (with permission from the author), shows P-wave velocity changes observed by 4D (time-dependent) tomography during the stimulation of the borehole GPK2 in the year 2000. A detailed discussion of the procedure used to observe and determine the P-wave changes is presented by Calo et al. (2012).

It is seen that the region of detected microseismicity (the cloud of black dots is small compared to the region where the P-wave velocity is reduced by as much as 20% in some regions). Some of the changes in velocity were temporary, suggesting that they may be related to temporal changes in fluid pressure; other changes appeared to be more permanent deformation that occurred aseismically.

These observations indicate that microseismicity, although a valuable indicator of the response of a rock mass to stimulation by fluid injection, does not identify the complete region influenced by a stimulation.

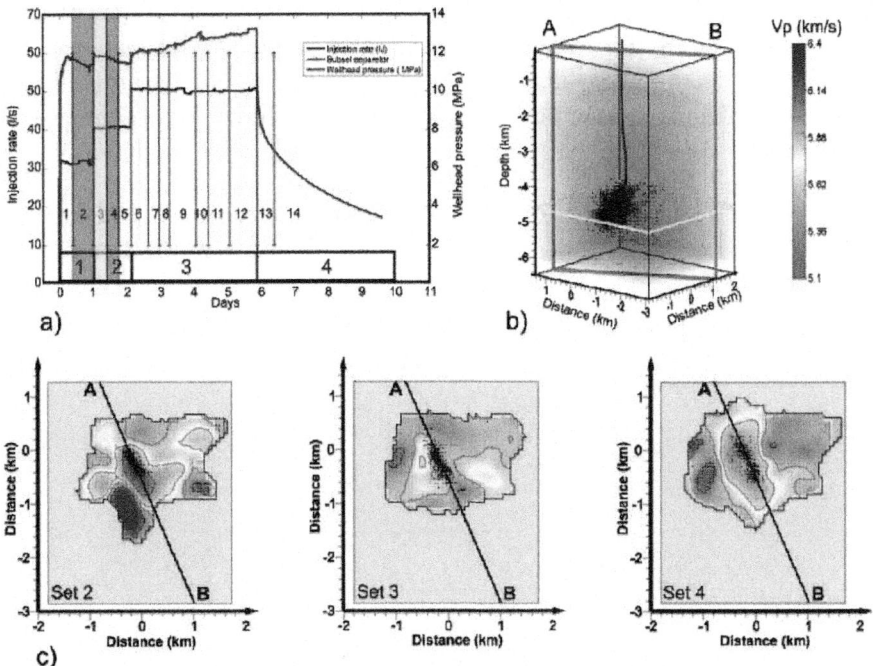

Figure 10: Aseismic slip induced by forced fluid flow as detected by P-wave tomography. (Soultz- sous- Fôrets, France. (a) The injection program (black curve is flow rate,

blue curve is well head pressure, horizontal axis is time in days); (b) 3D view of the seismic cloud with respect to the GPK2 borehole. Vertical axis is depth and horizontal axes are distances respectively toward the north and toward the east; and (c) horizontal projections corresponding to the yellow horizontal plane. The vertical green plane is shown as line AB in the plots of part c. P-wave velocity tomography for sets 2, 3 and 4 are indicated respectively by orange, yellow and green colors in the injection program. The vertical axis corresponds to North.

IN-SITU STRESS

As already noted, hydraulic fractures tend to develop in a more or less planar fashion, extending normal to the minimum regional principal stress. Determining the direction, and perhaps the magnitude, of the regional minimum stress is an important element of hydraulic fracturing strategy, especially with the development of directional drilling, which allows borehole to be drilled in the direction considered most favorable for fracturing with respect to stress direction. (see e.g., Figure 15 and related discussion).

Determination of the in-situ stress state also can be a significant challenge.

Stress in rock is distributed throughout the mass, and is influenced by the complicated structure of the mass[13] - . Most techniques of stress determination rely on what are essentially 'point' determinations. One difficulty of determining the regional stress is illustrated by the simple, albeit somewhat artificial, example of Figure 11. This shows a two-dimensional numerical model of the stress distribution in an elastic plate containing several finite frictional fractures.

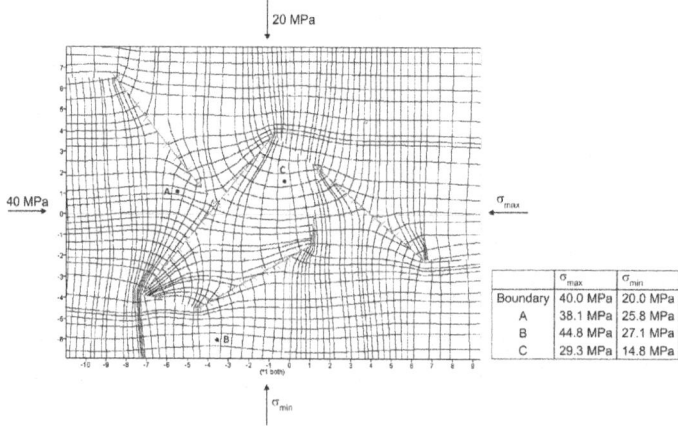

Figure 11: Influence of frictional cracks on the distribution and orientation of principal stresses, illustrative example.

The exercise serves to illustrate the difficulty of making stress determinations from local point measurements, be they in a borehole or on the surface. Stresses can change in orientation and magnitude locally due to geological inhomogeneities, fractures, faults, etc., many of which may be hidden or cannot be observed from the measurement location. Although determinations made at points A and B are reasonably close to the boundary values, point C is considerably different, and the directions of principal stress, as indicated by the principal stress trajectories, can be very different from the (regional) orientations, i.e., at the model boundary.

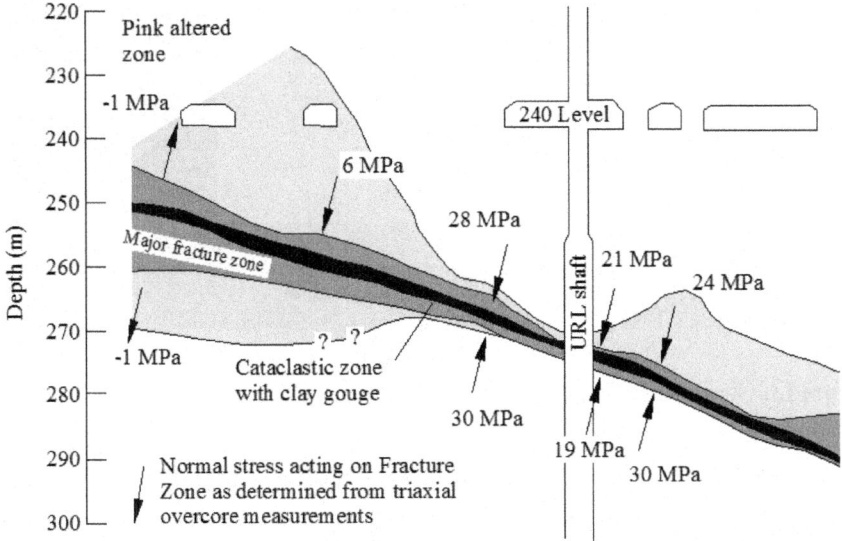

Observed Variability of Normal Stress Across a Thrust Fault Underground Research Laboratory Pinawa, Canada.

Figure 12: Normal stress variation across a thrust fault, Underground Research Laboratory, Canada.

Figure 12 provides an actual example of the variability of stress over relatively short distances. (The vertical and horizontal scales are equal in Figure 12). In this case, the main interest was to assess how normal stresses were affected by the thickness of gouge in the plane of the thrust fault.

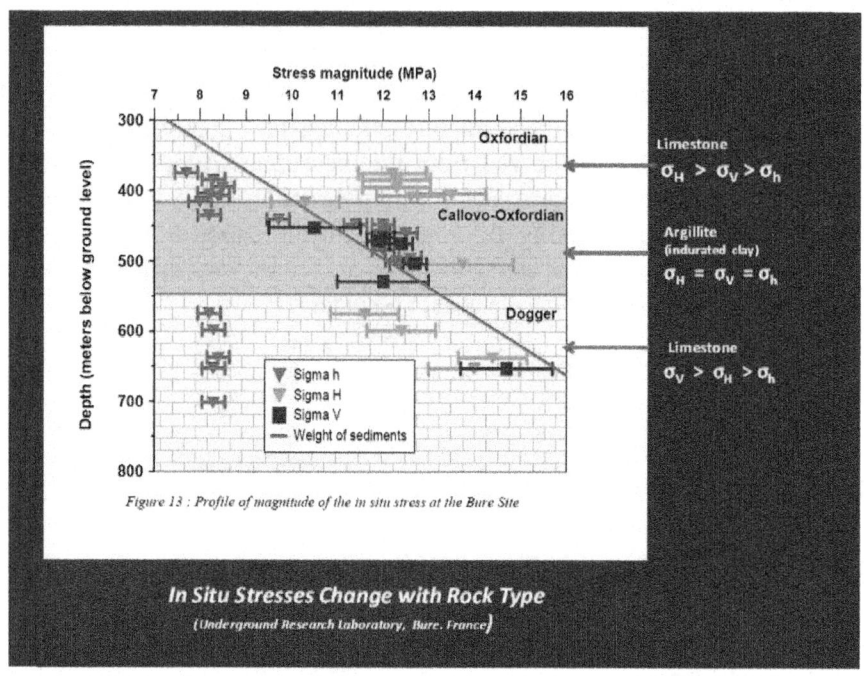

Figure 13 : Profile of magnitude of the in situ stress at the Bure Site

Figure 13: Observed stress distributions in argillite and limestones at the Underground Research Laboratory, Bure, France.

Figure 13 illustrates another important geological influence on stress distribution, changing lithology. This example is from the French Underground Research Laboratory (URL) [14] - at Bure in NE France. Laboratory tests on specimens of the Callovo-Oxfordien Argillite indicate a long-term viscosity of this rock suggesting that any imposed deviatoric stresses would tend towards an isotropic stress state over the order of 10 million years.

Test specimens from the limestones above and below the argillite do not appear to exhibit such viscosity. The stress distributions determined from field measurements support such differences in rheological characteristics of the rock formations.

Commenting on the in-situ stresses observations at Bure (i.e., as shown in Figure 13) Cornet (2012) notes as follows:

"Further, the complete absence of microseismicity in the Paris Basin (Grünthal and Wahlström, 2003, Fig. 4) and the absence of large scale horizontal motion as detected by GPS monitoring (Nocquet and Calais, 2004) indicate that no significant horizontal large-scale active deformation process exists today in this area.

"The important conclusion here is that the natural stress field measured on a 100 km² area at depth ranging between 300 m and 700 m does not vary linearly with depth and is not controlled by friction on preexisting well-oriented faults. Rather, the stress magnitudes seem to be controlled by the creeping characteristics of the various layers rather than by their elastic characteristics, with a loading mechanism that remains to be identified but which is neither related directly to gravity nor apparently to present tectonics.

"It is concluded here that the smoothing out of stress variations with depth into linear trends may be convenient for gross extrapolation to greater depth. But it should not be taken as a demonstration that vertical stress profiles in sedimentary rocks are governed by friction along optimally oriented faults, given the absence of both microseismicity and actively creeping fault. It should not be used for integrating together stress tensor components obtained within layers with different rheological characteristics."

Other examples could be cited, but the message is clear. Determination of in-situ stress in rock is an extremely challenging task, with results subject to considerable variability and uncertainty.

Stress orientations can be estimated from consideration of regional tectonics, faulting and interpretation of evidence from local structural geology supported in some cases by evidence based on borehole logs (e.g., tensile fractures induced along the well bore). Stress magnitudes are, in general, more difficult to determine and usually less significant, except as indicators of how stresses may be distributed across a site where the geology and engineering design are complex. In such cases, interpretation of stress distribution is best done in conjunction with a numerical model of the site, preferably one that includes the influence of important uncertainties and discussion with structural geologists familiar with the area under study.

CRITICAL STRESS STATE' IN THE EARTH'S CRUST

It is sometimes asserted that the Earth's crust is everywhere close to a 'critical state of stress,' i.e., that a small change in the devatoric stress in the rock is likely to produce slip on one or more faults with associated seismic activity. The current global interest in development of major resources of natural gas, the central role of hydraulic fracturing in this development, and the public apprehension that hydraulic fracturing will 'trigger earthquakes' has led to strong opposition to fracturing, and even legislation to ban the use of hydraulic fracturing in some countries and some States in the USA.

As illustrated by Figure 14, the seismic hazard, (i.e., probability of a damaging earthquake) varies very considerably from place to place. Thus, an earthquake of a given magnitude is 1000 times more likely to occur in Southern California than it is in the Eastern United States. The hazard is even lower in regions such as Texas, North Dakota and in the stable Canadian Shield region of the North American tectonic plate. While many earthquakes are initiated at depths considerably greater than depths where hydraulic fracturing is applied, it seems plausible to suggest that there may be less potential for fracturing to induce seismic activity in regions that have low seismic hazard. Also, as indicated by the comments of Cornet in the previous section of this paper, there is evidence that the critical stress hypothesis warrants detailed scrutiny, at least. This could have major implications for development of the world's major natural gas and EGS (enhanced geothermal systems) resources. Two recent studies,National Research Council (2012) and Royal Society – Royal Academy of Engineering (2012), have each concluded that the risk that hydraulic fracturing as used in development of energy resources would trigger significant seismic activity is small, but it would be valuable to examine the critical stress hypothesis more rigorously than has been done to date.

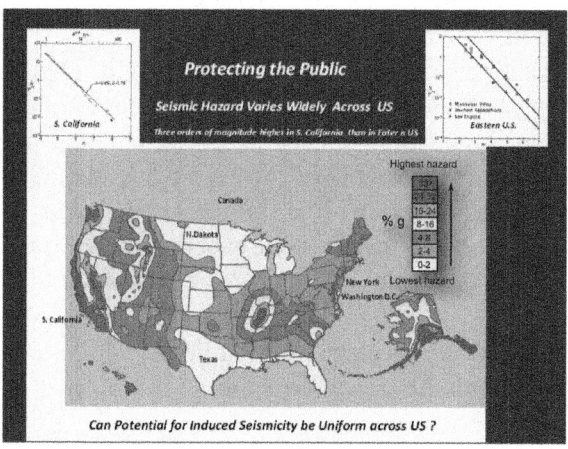

Figure 14: Seismic hazard map of the United States — US Geological Survey.

HYDRAULIC FRACTURING IN TIGHT SHALES

The development of inclined and horizontal drilling (see Appendix 1 - Figure A1-2) has helped stimulate intense activity to develop natural gas production from so-called tight shale, i.e., rock in which natural gas is held tightly within the very fine pore structure of the rock. Figure 15 illustrates the procedure used to stimulate these shales. The well is drilled horizontally in the gas-bearing formation, more or less in the direction of the minimum principal stress. Hydraulic fractures are generated (and propped) at intervals along the well to generate a network of connected flow paths that will allow the gas to flow to the well. Depth (i.e., extent) and spacing of the fractures should be optimized to produce the formations effectively. Bunger et al. (2012) discuss the factors in the design of an effective fracture strategy.

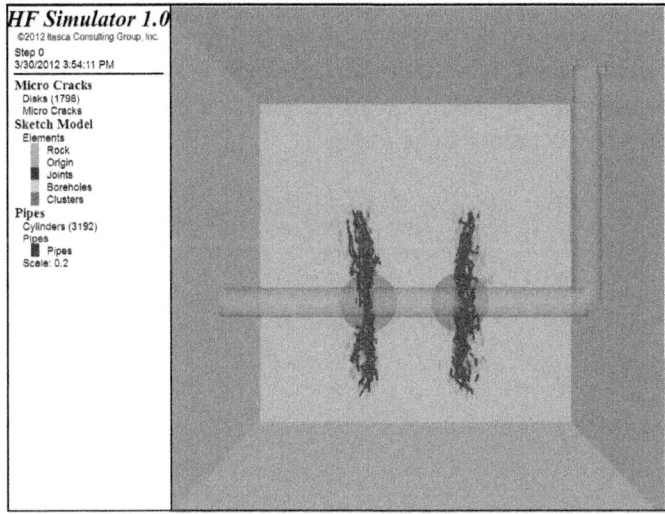

Figure 15: Staged hydraulic fracturing in a horizontal well. There may be many such wells along the horizontal well.

 ## Why Doesn't Microseismicity Correlate With Production?

The Total Rock Volume Affected by Microseismicity Accounts for Less Than 1% of Gas Production in First 6 Months

Figure 16: The volume of rock defined by microseismicity is a very small fraction of the volume producing gas.

Figure 16 shows a slide from a recent presentation by Prof. Mark Zoback, who kindly agreed to allow the author to include it here. Although on a somewhat smaller scale, the fact that considerable deformation and fracturing must be taking place that is not associated with detected microseismicity is similar to the phenomena discussed in connection with Figure 10. Prof. Zoback refers to such aseismic deformation as slow slip, and is conducting research to understand the underlying mechanisms, including the possible influence of the clay content of the shale. As can be seen in Figure 17 (courtesy of Prof. Zoback), the clay content can be large.

Average Shale Properties

	BARNETT	MARCELLUS	EAGLE FORD	FLOYD
Depth (ft)	3 - 9,000	2 - 9,500	4 - 13,500	6 - 13,000
TOC (%)	1 - 10	1 - 15	2 - 7	1 - 7
RO (%)	0.7 - 2.3	0.5 - 4+	0.5 - 1.7	0.7 - 2+
Porosity (%)	2 - 14	2 - 15	6 - 14	1 - 12
Qtz + Calcite (%)	40 - 50	40 - 60	50 - 80	20 - 30
Clay (%)	20 - 40	30 - 50	15 - 35	45 - 65
Areal Extent (mi²)	22,000	60,000	15,000	6,000
Resource Size (Tcf)	25 - 250	50 - 500	10 - 100	<<1

How many Floyd Shales are There?

Figure 17: Clay content of some typical 'tight' gas shales.

Figure 18 illustrates the very fine, micron scale, pore structure of a typical tight shale. Although the mechanism(s) by which flow pathways are established in such a fine structure is not clear, the level of microseismic energy release associated with brittle breakage of one or a few bonds will be very small and of high frequency (such that the radiated energy would be rapidly attenuated), and hence, not detectable by any geophone. Thus, absence of microseismicity may not indicate an absence of breakage of brittle bonds. Some mechanism must be operative that generates flow pathways. Intuitively, it might be expected that the clay content of the shale might lead to ductile and viscous deformation that could tend to close the pathways.

Figure 18: a) Outer surface of a FIB-SEM (Focused Ion Beam- Scanning Electron Microscope) volume of Eagle Ford Shale; (b) Transparency view of the distribution of connected pores (blue), isolated pores (red) and organic matter (green). (Courtesy of Prof. Amos Nur and J. Wallis (see Wallis et al., (2012) for details of technology.)

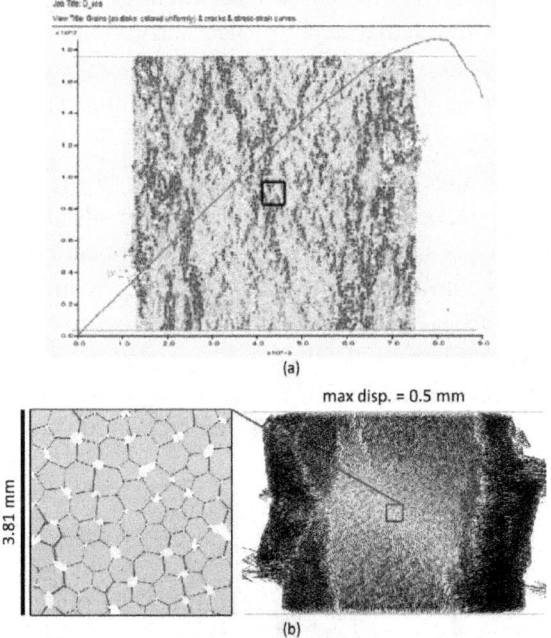

Figure 19: Micro-rupture of bonds within a *PFC* model of a rock loaded to failure, and beyond, in uniaxial compression. The darker red regions in (a) indicate coalescence of smaller groups of bonds that have ruptured. Eventually these larger regions develop to provide a mechanism that leads to collapse of the specimen. It is seen that bond breakage occurs throughout the specimen as the load is increased. The larger dark red regions will release larger amplitude, lower frequency waves that can be detected, whereas the smaller 'pathways' cannot be detected seismically. The load-deformation curve is shown as an 'overlay' on the specimen.

FRACTURE NETWORK ENGINEERING

This paper has emphasized the central role of fractures in rock, primarily natural fractures developed on a wide spectrum of scales over many tectonic epochs and many millions of years. These fractures and fracture systems are of special significance with respect to hydraulic fracturing and related techniques of fluid injection into rock since the fluid will tend to seek out those fractures that can be more readily opened against the local in-situ stress field as the fluid is injected. Given the complexity and lack of information on the fracture system, stress environment, etc., how can the engineering of hydraulic fracturing and related fluid injection programs advance most effectively?

Confronted with the same complexity of rock in situ, civil engineers and mining engineers have tended to adopt the 'Observational Approach' (Peck, 1969). In essence, this approach involves developing an initial engineering design for the problem, based on a first assessment/estimate of the rock (or soil) properties. Observe the actual performance and modify the initial design as needed to arrive at the desired performance. An example of the Observational Approach (as used in the New Austrian Tunnelling Method) is discussed in Fairhurst and Carranza-Torres (2002), see pp. 24-30.

Application of the Observational Approach to Hydraulic Fracturing and related fluid injection techniques faces some disadvantages and some advantages. We do not have 3D access to the engineering site. We do have powerful numerical modeling tools to help make a more informed initial estimate of how the system will perform; and we have sensing systems, both downhole and remote. Figure 20 illustrates a procedure that tries to apply the Observational Approach to hydraulic fracturing and related systems. The illustration describes an application to the extraction of Geothermal Energy.

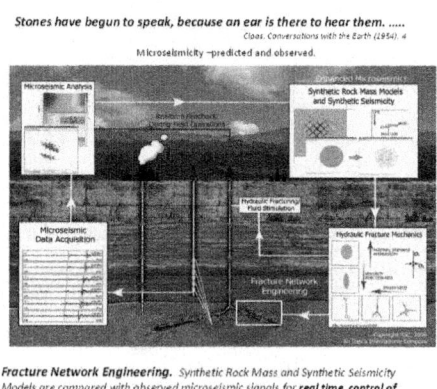

*Fracture Network Engineering. Synthetic Rock Mass and Synthetic Seismicity Models are compared with observed microseismic signals for **real time control of fracture network development.** (Enhanced Geothermal Systems.)*

Figure 20: Fracture network engineering system.

In this application, an initial design approach is developed based on a numerical modeling study incorporating any available data, insight, etc., on the site. This model provides an initial prediction of the performance. Instrumentation, both downhole and on-surface observes the initial response of the system and compares it with the prediction. This triggers a feedback signal to modify the design input to move the performance closer to the one desired. This iteration continues, changing progressively towards the performance desired.

Although the writer knows of no such Fracture Network Engineering system currently in operation, many of the components are available and it is time to start.

CONCLUSIONS

Expectations for higher living standards of a rising world population, and the associated demand for Earth's resources of energy, minerals and water, lead inevitably to greater focus on resources of the subsurface.

This focus includes the need to develop improved technology to develop these resources, and a better understanding of the nature of the subsurface environment as an engineering material.

Earthquakes and dynamic releases of energy are a daily reminder that on the global scale, Earth is critically stressed, and constantly trying to adjust seeking to achieve a condition of minimum potential energy for the entire system.

On going for many, many millions of years, such adjustments have resulted in the heterogeneous assembly of blocks of rock bounded by essentially planar surfaces; fault, fractures and similar 'discontinuities' varying in scale from tectonic plates and continents down to micron and even nanometers.

Some of these volumes are critically stressed; others are far from a critical condition. National maps of seismic hazards provide evidence of this heterogeneity on a larger scale.

Although Earth Resource Engineering activities may be kilometers in extent, they are small-scale within the larger Earth context. Subsurface engineering in a critically stressed region can be a much different challenge than in a stable region. It is important to assess the initial conditions carefully for each case, and especially where fluid injection is a main component of a project.

The sub-surface is opaque in several ways. Details of the key features that can control the response to an engineering activity in the sub-surface are often unknown. Problems are data-limited. This is particularly the case when the engineering is based on deep borehole systems, as in hydraulic fracturing and related fluid injection technologies.

Although operating in ways that may appear complex, the response of the subsurface to stimulation does obey the laws of Newtonian mechanics, and it is clear that pre-existing natural discontinuities have a major influence on how the subsurface responds to engineered changes.

The advent of powerful computers and developments in numerical modeling provide a potentially major tool to help develop better-informed strategies of subsurface engineering. Used interactively in close conjunction with instrumentation, both downhole and surface based, it should be possible to progressively develop a mechanics-informed understanding and path forward for more effective subsurface engineering.

Much as the field of Fracture Mechanics has led, and continues to lead, to major technological improvements for fabricated materials, so can development of the field of Rock Fracture Mechanics be of transformative value to subsurface engineering, and to society in general.

Hydraulic fracturing and related injection-stimulation systems will certainly be a central element in the future of Earth Resource Engineering. The organizers of HF 2013 are to be commended for focusing attention on this critically important topic.

APPENDIX 1

Earth Resources Engineering

In 2006, the US Academy of Engineering introduced the term 'Earth Resources Engineering' to replace 'Petroleum, Mining and Geological Engineering' in recognition of the broader range of engineering activities and concerns associated with use of the subsurface. The new title, it is hoped, will also stimulate important synergies between the various disciplines involved. Mining and civil engineers, for example, have direct three-dimensional access to the subsurface not available to colleagues in other subsurface activities. This access provides a major opportunity to conduct research and gain understanding of the mechanics of subsurface processes under actual in-situ conditions, as exemplified by Jeffrey et al. (2009), see Figure A1-1.

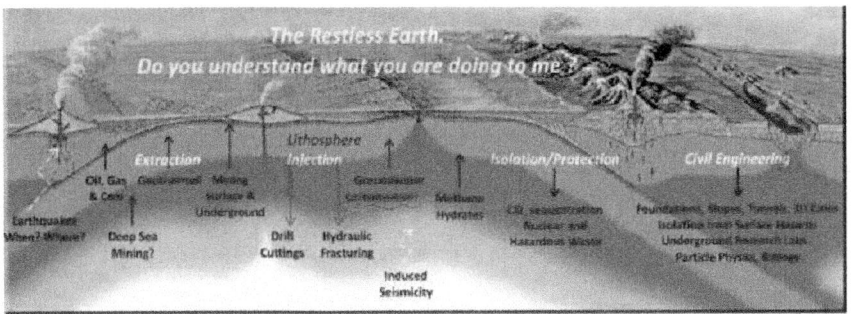

Figure A1-1: The restless Earth. Earth Resource Engineering activities are all confined to a very shallow part of the 40 km -700 km thick Earth's solid crust (lithosphere). Deepest borehole ~ 12 km; mine ~ 4km. Rock stress increases vertically σv ~
27MPa/km; laterally σh~ (0.5- 3.0).σv: Pore water pressure p = 10 MPa /km; temperature increase ~25°C /km depth.

Study of slip on active faults is a good example.

"The physics of earthquake processes has remained enigmatic due partly to
a lack of direct and near-field observations that are essential for the validation
of models and concepts. DAFSAM[15] -proposes to reduce significantly this
limitation by conducting research in deep mines that are unique laboratories
for full-scale analysis of seismogenic processes. The mines provide a 'missing
link' that bridges between the failure of simple and small samples in laboratory
experiments, and earthquakes along complex and large faults in the crust. There
is no practical way to conduct such analyses in other environment. To unravel
the complexity of earthquake processes, this project is designed as integrated
multidisciplinary studies of specialists from seismology, structural geology,
mining and rock engineering, geophysics, rock mechanics, geochemistry and
geobiology. The scientific objectives of the project are the characterization of
near-field behavior of active faults before, during and after earthquakes".[16]
- See also http://www.iris.edu/hq/instrumentation_meeting/files/pdfs/IRIS_
Johnston.pdf

Petroleum engineers can now reach depths in excess of 6 km and have
developed advanced drilling control technologies that allow precise access to
locations extending horizontally to more than 10-15 km from a single vertical
hole (see Figure 2).

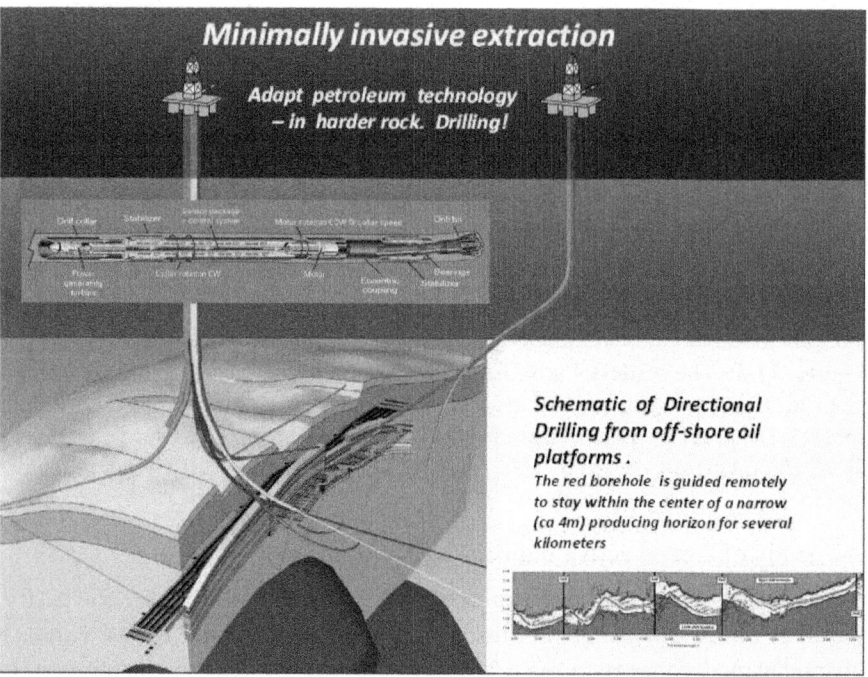

Figure A1-2: Schematic illustration of directional drilling for petroleum production.

These and related developments are stimulating interest in application of borehole technologies to other areas of subsurface engineering, including the development of less-invasive mining technologies, i.e., borehole extraction of minerals. Some applications, e.g., where crystalline rocks are involved, are contingent on the development of significantly lower-cost drilling technologies. The critical dependence of society on reliable and economic subsurface engineering is illustrated by the fact that currently more than 60% of the world's energy is delivered via a borehole. The Deepwater Horizon accident in the Gulf of Mexico in April 2010 provides a sober example of the consequences of error. In summary, hydraulic fracturing and related stimulation technologies are likely to see application to an increasing range of subsurface engineering challenges. HF2013, the first International Conference for Effective and Sustainable Hydraulic Fracturing, is very timely.

APPENDIX 2

Effect of Coring In Pre-Stressed Rock

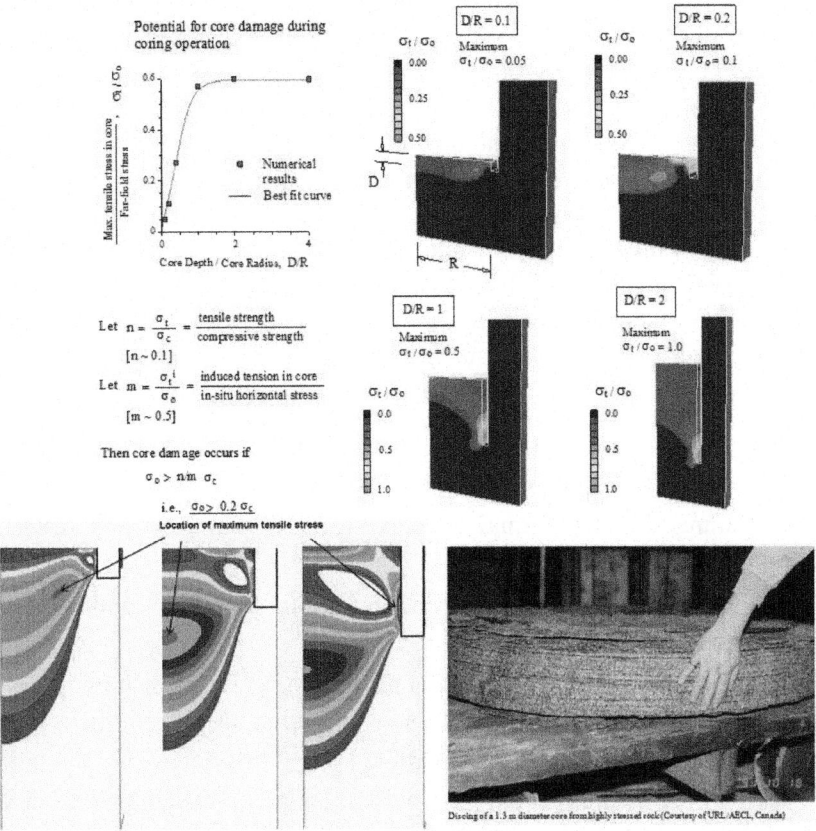

Figure A-2.1: Tensile stress concentrations induced in a brittle rock during coring.

The consequences of disturbing a pre-stressed rock medium are illustrated by examining the rock coring operation. Figure A2-1 shows the stress concentrations in a rock core in a brittle rock. If the in-situ stress normal to the axis of drilling is sufficiently high tensile cracks can develop in the core. Where lateral stresses are very high, then tensile 'spalling' may result, as shown in the photograph of the bottom right of Figure A2-1. Where the rock is more 'ductile' the core may undergo permanent deformation without fracturing. In both cases, the mechanical properties of these cores may differ significantly from those of the rock in situ from which the core was obtained.

ACKNOWLEDGEMENTS

Much of the material and concepts discussed in this paper is the result of work and discussions over many years with colleagues at Itasca Consulting Group, Inc. in Minneapolis and faculty in GeoEngineering at the University of Minnesota, especially in this instance, Professor Emmanuel Detournay. Particular help was received from Itasca colleagues Varun, Branko Damjanac, David Potyondy and Mark Lorig, The influence of numerous stimulating discussions with Professor François Cornet of the Institut de Physique du Globe, Strasbourg, France are clearly evident in the paper. Professors Amos Nur and Mark Zoback, of Stanford University, USA and of Ingrain, Inc., Houston, USA assisted with valuable material, as acknowledged in the text. Dr. Rob Jeffrey and Andrew Bunger of CSIRO, Melbourne, the leaders in arranging HF2013, have provided valuable comments, assistance and understanding throughout. To all, I am very grateful. Such invaluable assistance notwithstanding, I accept full responsibility for the interpretations and views expressed in the paper.

Notes

[1] - 1 exajoule =1018 joules = 1018 watt.seconds.

[2] - Future of Geothermal Energy (2005) Synopsis and Executive Summaryp.1-4 (2).

[3] - Future of Geothermal Energy (2005)Synopsis and Executive Summaryp.1-5 (5).

[4] - A fractured rock mass is typically about two orders of magnitude lower in strength than the strength of a laboratory specimen taken from the rock mass [Cundall (2008); Cundall et al, (2008)].

[5] - Hydraulic fractures generated in classical petroleum applications typically extend (2b) of the order of 25m ~ 50m from a wellbore. The fracture aperture (2a) at the wellbore then will be typically of the order of 0.01 m. Thus, the tensile stress concentration at the tip is very high of the order of 103.

[6] - In his second paper, Griffith (1924), demonstrated that tensile stresses also developed around similar cracks loaded in compression, provided the cracks were inclined to the direction of the major principal (compressive) stress.(He also assumed that the cracks did not close under the compression.) For the optimum crack inclination, an applied compressive stress of eight times the magnitude of the tensile strength was required to develop a tensile stress on the crack boundary (close to, but not at the apex of the crack) equal to the limiting value in the tensile test. He concluded that the uniaxial compressive strength of a brittle material should be eight times greater than the tensile strength. Interestingly, he did not invoke his second (minimum potential energy)

criterion. It was later determined that although a tensile crack could initiate in a compressive stress regime as predicted by Griffith (1924), the crack was stable (i.e., did not satisfy the minimum potential energy criterion). The compressive/ tensile strength ratio is greater than 8 (see Hoek and Bieniawski, 1966).

[7] - Tension is assumed to be positive in Figure 3.

[8] - A typical hydraulic fracture may have a length (2a) of the order of 50m and a maximum aperture (2b) of 5mm, so that the stress concentration will be of the order of 2000:1.

[9] - Commonwealth Scientific and Industrial Research Organization.

[10] - Typically, computer tests indicate the unconfined strength of a Synthetic Rock Mass of the order of 50-m to 100-m side length, to be a few percent of the unconfined strength of the laboratory specimen.

[11] - A number of important subsurface engineering problems involve borehole access only. This often means difficulty in establishing reliable, realistic DFN's. In such cases there is no recourse, at least at the start of the project, other than to try to infer fracture networks from borehole observations, perhaps supplemented by local observations of structural geological features . The DFN for Northparkes was available and convenient to use in the example shown in Figure 5.

[12] - Adapted from Fairhurst (1971).

[13] - See also footnote 17 –Appendix 1.

[14] - The URL at Bure was developed in order to determine the suitability of the Calllovo-Oxfordien Argillite formation for permanent storage of high– level nuclear waste.

[15] - DAFSAM -Drilling Active Faults in South African Mines.

[16] - http://www.icdp-online.org/front_content.php?idcat=460

REFERENCES

1.   T. L Anderson, 2005Fracture Mechanics: Fundamentals and Applications. 3rd edition, CRC Press (0-84931-656-1

2.   E. V Artyushkov, 1973Stresses in the Lithosphere Caused by Crustal Thickness

3.   J Inhomogeneities, Geophy.Res. 7832November 10, 1973

4.   A. P Bunger, X Zhang, and R. G Jeffrey, 2012Parameters Affecting the Interaction Among Closely Spaced Hydraulic Fractures" SPE Journal March 2012, 292306

5. M Calo, C Dorbath, F. H Cornet, and N Cuenot, 2011Large scale aseismic motion identified through 4D P-wave tomography; Geophys. J. Int. 18612951314

6. P. A Cundall, 2008An Approach to Rock Mass Modelling," in From Rock Mass to Rock Model-CD Workshop Presentations (15 September, 2008)-SHIRMS 2008 (Proc. 1st Southern Hemisphere International Rock Symposium, Perth, Western Australia, September 2008) Y. Potvin et al., Eds. Nedlands, Western Australia: Australian Centre for Geomechanics.

7. T. T Cladouhos, M Clyne, M Nichols, S Petty, W. L Osborn, and L Nofziger, 2011Newberry Volcano EGS Demonstration Stimulation Modeling" GRC Transactions, 35317322

8. F. H Cornet, 2012The relationship between seismic and aseismic motions induced by forced fluid injections." Hydrogeology Journal (2012) 20: 1463-1466

9. F. H Cornet, and T Röckel, 2012Vertical stress profiles and the significance of "stress decoupling". Tectonophysics 5812012193205

10. P. A Cundall, M. E Pierce, and D. Mas Ivars. (2008Quantifying the Size Effect of Rock Mass Strength" in SHIRMS 2008 (op.cit.) 2315

11. B Damjanac, and C Fairhurst, 2010Evidence for a Long-Term Strength Threshold in Crystalline Rock,‖ Rock Mech. Rock Eng., 43, 513-531 (2010).

12. B Damjanac, C Detournay, P. A Cundall, and Varun, (2013Three-Dimensional Numerical Model of Hydraulic Fracturing in Fractured Rock Masses" Proc. HF 2013The International Conference for Effective and Sustainable Hydraulic Fracturing, Brisbane, May 20-22, 2013

13. B Damjanac, and C Fairhurst, Evidence for a Long-Term Strength Threshold in Crystalline Rock,‖ Rock Mech. Rock Eng., 43, 513-531 (2010Duchane, D and D. Brown, (2002) "Hot Dry Rock (HDR) Geothermal Energy Research and Development at Fenton Hill, New Mexico" GHC (Geo-Heat Center) Bulletin, December. 2002 1319

14. C Fairhurst, and C Carranza-torres, 2002Closing the Circle- Some Comments on Design Procedures for Tunnel Supports in Rock," in Proceedings of the University of Minnesota 50th Annual Geotechnical Conference (February 2002), 2184J. F. Labuz and J. G. Bentler, Eds. Minneapolis: University of Minnesota, 2002. [available at www.itascacg. comgo to 'About'and Fairhurst Files]

15. C Fairhurst, 1971Fundamental Considerations Relating to the Strength of Rock. Colloquium on Rock Fracture, Ruhr University, Bochum, Germany, April 1971. (see http://www.itascacg.com/about/ff.php)Revised and

published in Report of the Workshop on Extreme Ground Motions at Yucca Mountain, August 23-25, 2004, U.S. Geological Survey, USGS Open-File Report 20061277T. C. Hanks et al., Eds. Reston, Virginia: USGS, 2006.

16. J Geertsma, and F De Klerk, 1969A Rapid Method of Predicting Width and Extent of Hydraulic Induced Fractures. J Pet Technol 211215711581SPE-2458-PA. http://dx.doi.org/10.2118/2458-PA

17. J. F Geyer, and S Nemat-nasser, 1982Experimental Investigation of Thermally induced Interacting Cracks in Brittle Solids Int. J. Solids and Structures 184349356

18. A. A Griffith, 1921The Phenomena of Rupture and Flow in Solids Phil. Trans. R. Soc. Lond. A 1921,, 221, 163-198 doi:rsta.1921.0006

19. A. A Griffith, 1924Theory of Rupture. Proc. First Int. Cong. Applied Mech (eds Bienzo and Burgers). 5563Delft: Technische Boekhandel and Drukkerij. 1924

20. G Grünthal, and R Wahlström, 2003An Mw-Based Earthquake Catalogue for Central, Northern and Northwestern Europe using a Hierarchy of Magnitude Conversions. J. Seismol. 7, 507-531 (Available at http://seismohazard.gfzpotsdam.de/projects/en/eq_cat/menue_e"q_cat_e.html)

21. E Hoek, and Z. T Bieniawski, 1966Fracture Propagation Mechanism in Hard Rock," in Proceedings of the First Congress of the International Society of Rock Mechanics. Lisbon, September-October, 1243249J. G. Zeitlen, Ed. Lisbon: LNEC.

22. E Hoek, and E. T Brown, 1980Underground Excavations in Rock." Inst'n of Mining and Metallurgy (London) Revised 1982, 164

23. G. C Howard, and C. R Fast, 1970Hydraulic Fracturing" SPE Monograph 2. Henry L.Doherty Series 203 pp. SPE 30402

24. C. E Inglis, 1913Stresses in a Plate Due to the Presence of Cracks and Sharp Corners," Trans. Inst. Naval Arch., London, 55(1), 219141

25. R. G Jeffrey, et al2009Measuring Hydraulic Fracture Growth in Naturally Fractured Rock. SPE 124919; SPE Annual Technical Conference and Exhibition, New Orleans, Louisiana, USA, 47October 2009

26. J. F Knott, 1973Fundamentals of fracture mechanics, Wiley (0-47049-565-0

27. National Research Council2012Induced Seismicity Potential in Energy Technologies." Washington, DC: The National Academies Press, 2012. (300p.) View online at http://www.nap.edu/catalog.php?record_id=13355

28. J. M Nocquet, and E Calais, 2004Geodetic Measurements of Crustal Deformation in the Western Mediterranean and Europe " Pure Appl. Geophy., 161; 661668

29. R. P Nordgren, 1972Propagation of a Vertical Hydraulic Fracture. SPE J. 124306314SPE-3009-PA. http://dx.doi.org/10.2118/3009-PA.

30. R. B Peck, 1969Advantages and limitations of the observational method in applied soil mechanics. Geotechnique, 192171187

31. T. K Perkins, and L. R Kern, 1961Widths of Hydraulic Fractures. J Pet Technol 139937949SPE-89-PA. http://dx.doi.org/10.2118/89-PA.

32. W. S Pettitt, J. F Hazzard, B Damjanac, Y Han, M Pierce, T Katsaga, and P. A Cundall, Microseismic Imaging and Hydrofracture Numerical Simulations," in Proceedings, 21st Canadian Rock Mechanics Symposium (Alberta, Canada, May 5-9, 2012

33. M Pierce, 2011Discrete Fracture Network Simulation" DFN training session LOP (Large Open Pit). [ppt slides available on request. Itasca Consulting Group: www.itascacg.com]

34. A Riahi, and B Damjanac, 2013Numerical Study of Interaction between Hydraulic Fractures and Discrete Fracture Networks" Proc. HF 2013The International Conference for Effective and Sustainable Hydraulic Fracturing, Brisbane, May 20-22, 2013

35. Royal Society and Royal Academy of Engineering (2012Junep. "Shale Gas Extraction in the UK: a review of hydraulic fracturing" Issued: June 2012, DES2597] View report online at: royalsociety.org/policy/projects/shale-gas-extraction and raeng.org.uk/shale

36. O Scotti, and F. H Cornet, 1994In-Situ Evidence for Fluid-Induced Asesismic Slip Events along Fault Zones. Int. J. Rock Mech Min.Sci. &Geomech. Abstr. 314347258Control in Mines (1965). South African Institute of Mining and Metallurgy, Johannesburg, 606p

37. A. M Starfield, and P. A Cundall, 1988Towards a Methodology for Rock Mechanics Modelling" Int. J. Rock Mech. Min. Sci.& Geomech. Abstr. 25 (3) 99106

38. J. F Tester, et al2006The Future of Geothermal Energy"-Impact of Enhanced Geothermal

39. Systems (EGS) on the United States in the 21st CenturyMIT Press.

40. J. D Wallis, J Devito, and E Diaz, 2012Digital Rock Physics- A New Approach to Shale Reservoir Evaluation" Oilfield Technology, March 2012   [http://www.ingrainrocks.com/articles/a-new-approach-to-shale-reservoir-evaluation/]

41. X Zhang, and R. G Jeffrey, 2008Re-initiation or termination of fluid-driven fractures at frictional bedding interfaces" JGR, 113BO 8416, doi:10.1029/2007JB005327,

Chapter 7

ESTIMATING HYDRAULIC CONDUCTIVITY OF HIGHLY DISTURBED CLASTIC ROCKS IN TAIWAN

Cheng-Yu Ku[1] and Shih-Meng Hsu[2]
[1]National Taiwan Ocean University
[2]Sinotech Engineering Consultants, Inc Taiwan

INTRODUCTION

Understanding groundwater flow in fractured consolidated media has long been important when undertaking engineering tasks such as dam construction, mine development, the abstraction of petroleum, slope stabilization, and the construction of foundations. To study groundwater flow in support of these tasks, the focus of most hydrogeological investigations has been on the characterization of the hydraulic properties of the higherpermeability fractures in the rock mass.

Taiwan is situated on the edge of the Eurasian and Philippine Sea plate. Plate tectonics have created numerous fault lines that crisscross the island. As a result of high density of faults, rock core data with fractures, soft and cohesive gouges, and various lithologies are extensive in boreholes. In general, the permeability of clay-rich gouges has extremely low values. On the contrary, the fractures often have higher permeability. The hydraulic properties of fractured rocks in Taiwan, therefore, vary with highly disturbed geological structures and lithology. To obtain hydraulic properties of fractured rocks in Taiwan, the investigation of vertical variation of the fractures in a borehole is of importance. This study utilized a highresolution BoreHole acoustic TeleViewer (BHTV, Williams and Johnson, 2004) to scan images of the borehole. The information gathered from BHTV was used to characterize lithology and fractures for the borehole and was essential to conduct a proper measurement of rock mass hydraulic conductivity. The double packer systems were then used to determine the hydraulic conductivity in a portion of borehole using two inflatable packers. Although this type of test can directly measure the hydraulic parameter, costs of the testing are fairly high. Several studies (Black, 1987; Carlson and Olsson,1977; Louis, 1974; Burgess, 1977; Wei et al., 1995, Zhao, 1998) have proposed the estimation of rock mass hydraulic conductivity using different empirical equations. These empirical equations provide a great feature for characterizing rock mass hydraulic properties quickly and easily. However, the applicability of these equations in highly disturbed clastic sedimentary rocks in Taiwan is very limited.

This study proposed the establishment of an empirical HC model for estimating rock mass hydraulic conductivity of highly disturbed clastic sedimentary rocks in Taiwan using the BHTV and the double packer hydraulic tests. Four geological parameters including rock quality designation (RQD), depth index (DI), gouge content designation (GCD), and lithology permeability index (LPI) were adopted for establishing the empirical HC model. To verify rationality of the proposed HC model, 22 in-situ hydraulic tests were carried out to measure the hydraulic conductivity of the highly disturbed clastic sedimentary rocks in three boreholes at two different locations in Taiwan. Besides, the model verification using another borehole data with four additional in-situ hydraulic tests from similar clastic sedimentary rocks was also conducted to further verify the feasibility of the proposed empirical HC model. This paper presents the measured hydraulic conductivity results and the relationship among the hydraulic conductivity, RQD, DI, GCD, and LPI. The application of the proposed HC model was also addressed.

DESCRIPTION OF STUDY AREAS AND BOREHOLES

Description of Study Areas

Taiwan's strata are distributed in long and narrow strips, almost parallel to the island's axis. Metamorphic rock lies under the Central and Snow Mountain Ranges. Sedimentary rock forms part of the island-wide piedmonts and coastal plains as well as the Coastal Mountain Range. The island of Taiwan has three geological zones divided by longitudinal faults: the Central Range, Western Piedmont and Eastern Coastal Mountain Range zones (Fig. 1(a)). About 26 hydraulic conductivity measurements were conducted in four boreholes in Western Piedmont, primarily at three sites: Da-Keng, Shang-Ming, and Caoling (Fig. 1(a)) in which borehole HB-94-01 is in the Da-Keng site, boreholes HB-95-01 and HB-95-02 are in the Shang-Ming site, and borehole CH-04 is in the

Caoling site. Besides, the Da-Keng and Caoling sites are in central Taiwan and the Shang-Ming site is in southern Taiwan. The dominant rock strata of the Shang-Ming site include Miocene sedimentary rock with layers of sandstone or shale or their alternation. The major structures consist of a series of parallel easterly inclined thrust faults and folds, which often form local fractured zones, including geological structures such as the Pingshi fault, the Biauhu fault, and the Chin-Shan fault. Figure 1(b) presents the distribution of these geological strata and structures. In addition, borehole HB-94-01 in the Da-Keng site and borehole CH-04 in the Caoling site also have similar rock strata but without geological structures.

Based on the loggings and geological analysis, HB-95-01 and HB-95-02 are strongly influenced by the faults; nevertheless, HB-94-01 and CH-04 are not.

Boreholes

The depth of the borehole HB-94-01 is 110 m. The principal lithologic units for the borehole are sandstone and siltstone. The interval of 36 m to 44 m is a fractured zone compared to other depths in the borehole. A total of 8 hydraulic tests using a double packer system were carried out to determine hydraulic conductivity (Sinotech, 2006). The strategy of the test design from the drilling work was to determine hydraulic properties from different geological structures such as no fracture, a single fracture, or multiple fractures at different depths.

The drilling depths of HB-95-01 and HB-95-02 are 250 m and 350 m, respectively. The principal lithologic units for HB-95-01 are sandstone, argillaceous sandstone, and sandy mudstone. The principal lithologic units for HB-95-02 are sandstone, argillaceous sandstone, and sandstone mixed with some mudstone. HB-95-01 and HB-95-02 are close to the Biauhu fault and Pingshi fault, respectively (Fig. 1(b)). Rock core photos (Fig. 2(a)) indicated soft and cohesive gouges are extensive in both boreholes in which the hydraulic properties of fault-related rocks can be studied. The study completed 3 and 14 hydraulic tests in HB-95-01 and HB-95-02, respectively (Sinotech, 2006). The strategy of the test design was to determine hydraulic conductivity in more permeable zones and clay-rich gouge zones. Besides, the borehole CH-04 is not influenced by the faults (Fig. 2(b)) and used for the model verification and it is described in Section 5.7.

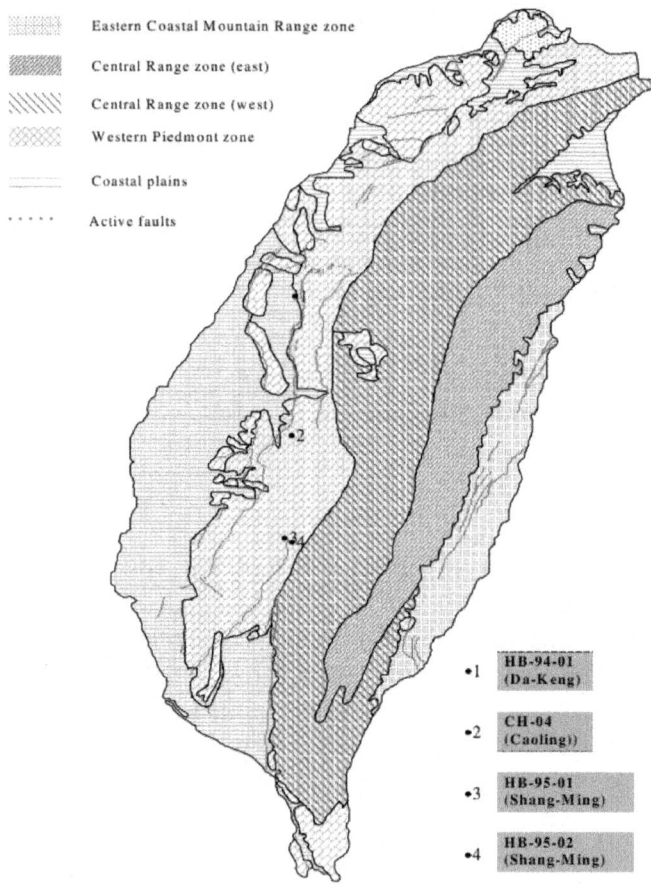

Figure 1. (a) Location of major faults and four boreholes for this study in Taiwan.

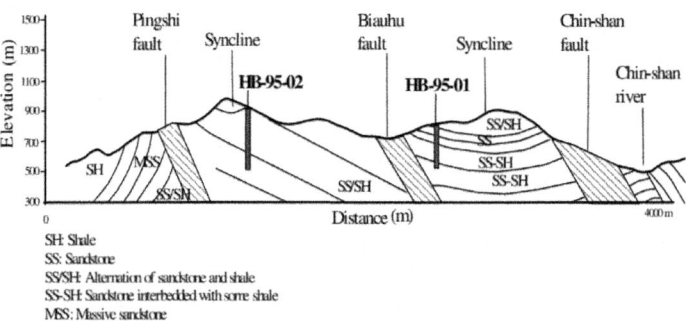

Figure 1. (b) Detailed distribution of geological strata and structures of boreholes HB-95-01 and HB-95-02.

Figure 2. (a) Rock core photos of borehole HB-95-2 with fault influence.

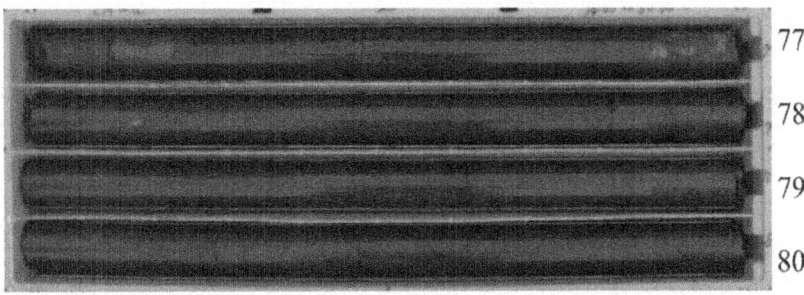

Figure 2. (b) Rock core photos of borehole CH-04 without fault influence.

HYDRAULIC CONDUCTIVITY OF FRACTURED ROCK MASSES

It is widely recognized that fracturing plays a decisive role in rock hydraulics, especially in low permeability rocks such as crystalline, volcanic and carbonate rocks, and in some classic sedimentary formations, such as sandstones, shales, glacial tills and clays. In highly disturbed fractured rocks, hydraulic properties depend on density, size, infillings and interconnection of fractures.

A distinction can be made between hydraulic conductivity of fracture and of intergranular (matrix) material. Previous study (Lee and Farmer, 1993) has proposed the hydraulic conductivity of a rock mass with three orthogonal joint sets with similar spacing ad constant aperture in all directions. The effect of stress on permeability is also of importance for estimating rock mass hydraulic conductivity (Snow, 1969). Several studies, shown in Table 1, have also pointed out that rock mass permeability may decrease systematically with depth (Black, 1987; Carlson and Olsson,1977; Louis, 1974; Burgess, 1977; Wei et al., 1995, Zhao, 1998). The decrease in permeability with depth in fractured rocks is usually attributed to reduction in fracture aperture and fracture spacing. The reduction is due to the effect of geostatic stresses, and thereby the permeability of fractured rocks will be reduced. Accordingly, the depth may be considered as a factor in evaluating rock mass permeability.

MEASUREMENT OF ROCK MASS HYDRAULIC CONDUCTIVITY

For decades, the determination of hydraulic properties in fracture rocks has been qualitatively estimated using the Lugeon test. It is now recognized that this approach is not suitable in highly disturbed fractured rocks. The type of test only gives an average value of hydraulic conductivity in a stratum and is not able to identify (1) aquifer's type in a required testing section; (2) storativity of an aquifer; and (3) relations between hydraulic properties and geological structures such as water-bearing fractured zones. Results from the test may be insufficient to characterize hydraulic properties for complex geological environments. They may be subject to hazards such as extensive water inflow during underground excavation.

To provide a better characterization of hydraulic properties of fractured rocks, a double packer technique can be adopted and is often utilized to overcome the shortcomings of the Lugeon test. Packers can be used to isolate a portion of borehole for hydraulic testing. Hydraulic properties for a single of fracture, a group of fractures, or an entire rock formation can be easily identified by the technique.

Table 1. Diverse approximations for estimating rock mass hydraulic conductivity.

Equation	Reference
$k = az^{-b}$	Black (1987) a and b are constants, z is the vertical depth below the groundwater surface.
$\log K = -8.9 - 1.671 \log Z$	Snow (1969) K (ft^2) is the permeability. z (ft) is the depth.
$K = 10^{-(1.6 \log z + 4)}$	Carlson and Olsson (1977) K (m/s) is the hydraulic conductivity. z (m) is the depth.
$K = K_s e^{(-Ah)}$	Louis (1974) K (m/s) is the hydraulic conductivity. K_s is the hydraulic conductivity near ground surface. h (m) is the depth. A is the hydraulic gradient.
$\log K = 5.57 + 0.352 \log Z$ $-0.978(\log Z)^2 + 0.167(\log Z)^3$	Burgess (1977) K (m/s) is the hydraulic conductivity. Z (m) is the depth.
$K = K_i[1 - Z / (58.0 + 1.02Z)]^3$	Wei et al. (1995) Z is the depth. K is the hydraulic conductivity. K_s (m/s) is the hydraulic conductivity near ground surface.

Borehole Acoustic Televiewer(BHTV) Investigation And Hydraulic Test Design

Prior to hydraulic testing, the study utilized a high-resolution borehole acoustic televiewer (BHTV) to scan images of boreholes. The information (Fig. 3(a) and (b)) gathered from BHTV was used to characterize lithology and fractures for the borehole and was essential to the proper design of the hydrogeological program. Test design is dependent on the characteristics of the zone tested and the desired information. Accordingly, the main testing strategy in this study was to detect waterbearing fractures. In addition, the study investigated the vertical variation of the hydraulic conductivity in a borehole and hydraulic property of fault-related rocks.

A water-bearing zone of subsurface commonly appears in the section with multiple fractures. According to BHTV logs from boreholes, the study selected locations with images that show multiple fractures as hydraulic test sections. Figure 3 shows that two test zones were selected by this strategy. Other testing zones for other study purposes can be selected by BHTV images.

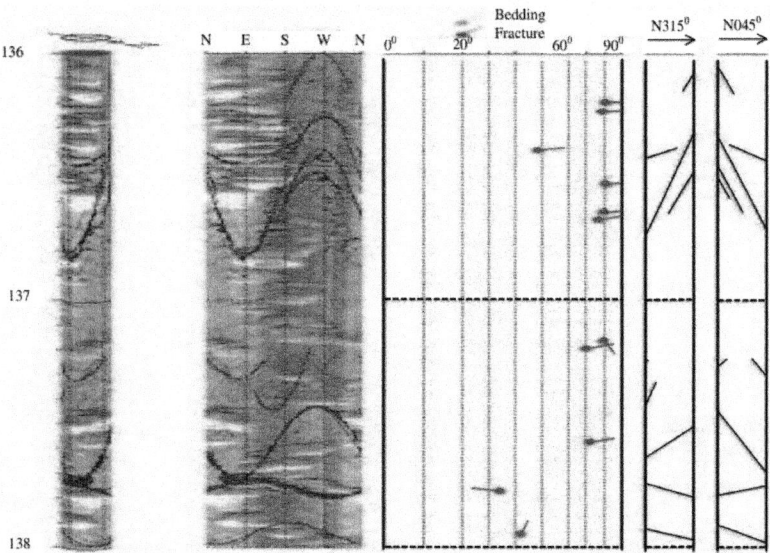

Figure 3. (a) The pack-off zones and their corresponding BHTV images (depth 136m~138m, HB-95-01).

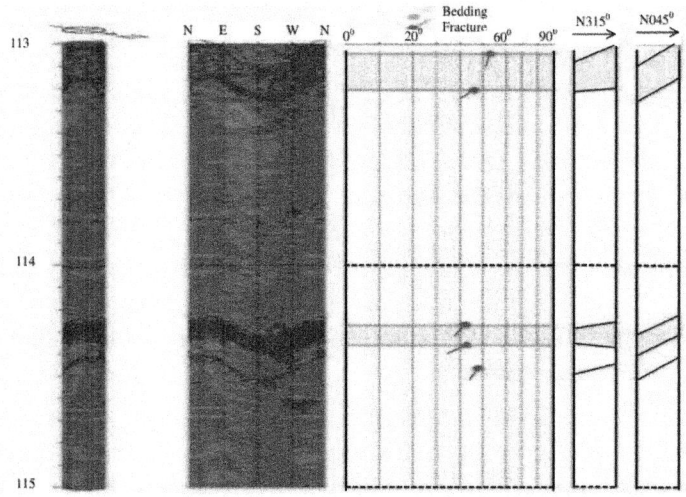

Figure 3. (b) Identification of shear band from BHTV (depth 113m~115m, HB-95-02).

Double Packer System

Double packer systems are the most commonly used tools for hydrogeological testing in boreholes. They can be used to determine the hydraulic property in a section of borehole based on two inflatable packers. It is now recognized that this approach is appropriate to investigate the variability of a borehole as it intersects various hydrogeological units. The double packer system in this study (Fig. 4) consists of two inflatable rubber packers, a shut-in valve, a submersible pump, and pressure transducers for monitoring above and below the packers and the isolated interval. The shut-in valve can open or close the hydraulic connection between the pipe string and the test section. The rubber packers can be inflated using nitrogen delivered through a polyethylene air line. The pumping or injecting rate can be monitored at the land surface with a flow meter. To conduct each hydraulic test, the packers are inflated to isolate a section of borehole, and the rate of flow and/or pressure in the test interval over a period of time can be measured.

The BHTV images for different test intervals in Boreholes HB-95-01 and HB-95-02 are shown in Fig. 5(a) and 5(b). It is obvious that the fractures can be identified clearly using our highresolution BHTV. The intervals in the depth from 118.5 m to 121.7 m and from 134.8 m to 138.0 m were sealed by double packers for conducting a pressure pulse test and a constant head injection test, respectively. Figures 5(a) and 5(b) show the results of hydraulic tests which were conducted by different type of hydraulic tests by means of AQTESOLVE. The type of hydraulic test chosen in this study for each test interval was decided by a hydraulic diagnosis test which mainly detects permeability of the test interval prior to a normal test. For the test interval of 118.5 m to 121.7 m, although three fractures and a fracture zone of approximately 7.25 cm thickness were seen on the borehole image, lack of interconnectivity of fractures and soft and cohesive gouges existing at the fractures may reduce the permeability of rock masses. In addition, four types of tests, including pumping tests, injection tests, slug tests, and pressure pulse tests can be applied to the double packer system. Pumping tests involve pumping at a constant or variable rate and measuring changes in water levels during pumping. In injection tests, fluid is injected into a test interval while keeping the head of the test interval at a constant value. A slug test involves the abrupt removal, addition, or displacement of a known volume of water and the subsequent monitoring of changes in water level as equilibrium conditions return. In a pressure pulse test, an increment of pressure is applied to a packed zone. The pressure decay is monitored. Typically, the decision on which type of test to perform is based on the expected permeability of the test interval, the volume of rock to be sampled, and the availability of time and equipment (NRC, 1996). Hydraulic properties determined by slug tests or pressure pulse tests are representative only for the material in the immediate vicinity of the borehole.

To obtain hydraulic conductivity over a large area, the procedure of a single-hole hydraulic test is to perform a pumping test at a test interval first. If the pumping test cannot be performed due to low permeability of the test section, a constant head injection test will be conducted instead. Once the flow rate cannot be measured by limitation of the flow meter (less than 0.11 l/min) during the injection test, a slug test or pressure pulse test can be performed. The duration of a pressure pulse test is much shorter than that of a slug test. For this reason, the pressure pulse test is commonly applied to test intervals of very low permeability. However, the volume of rock tested by a pressure pulse test is significantly smaller when compared to a slug test.

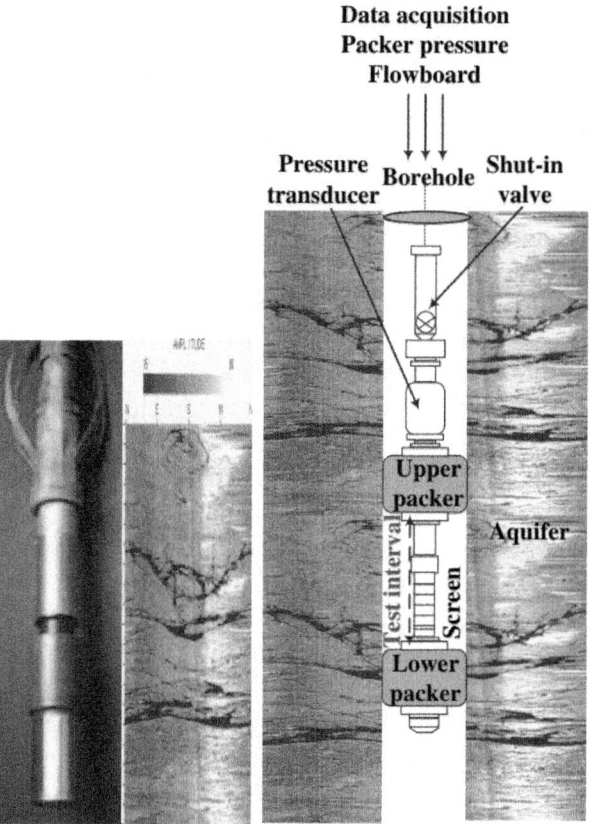

Figure 4. Schematic drawing of BHTV, acoustic image of borehole, and the double packer system.

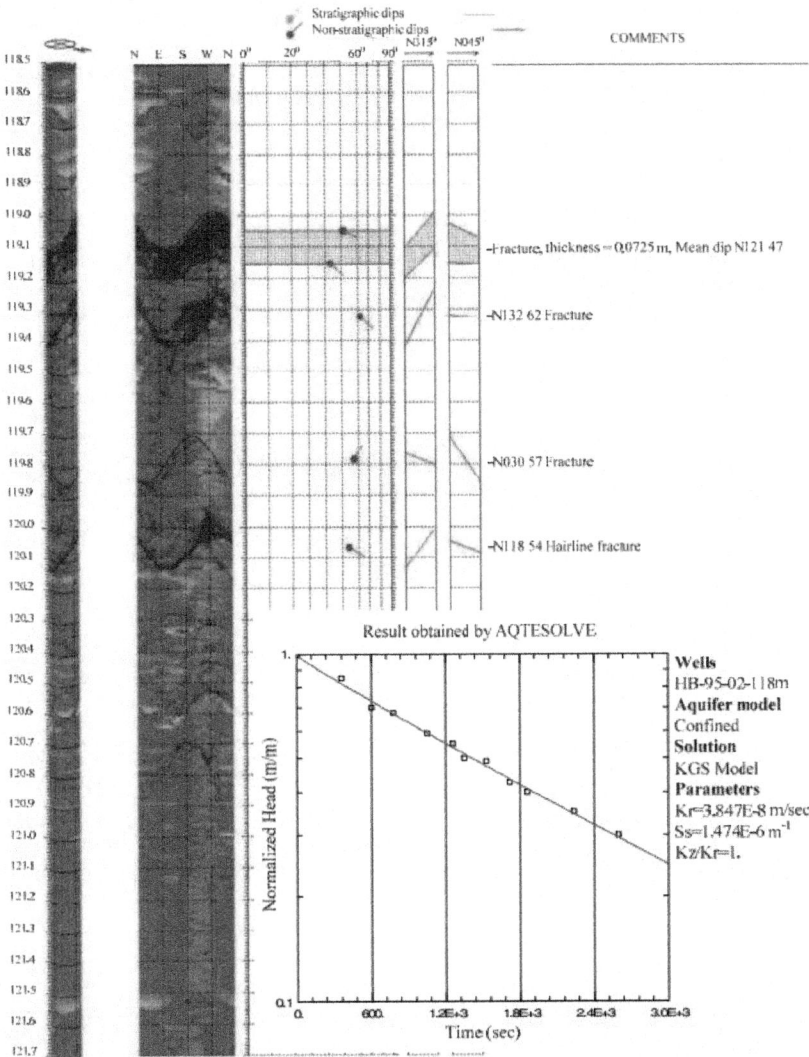

Figure 5. (a) Evaluation of hydraulic parameters using AQTESOLVE (right lower figure) and BHTV images at pack-off zones 118.5 m to 121.7 m in Borehole HB-95-02.

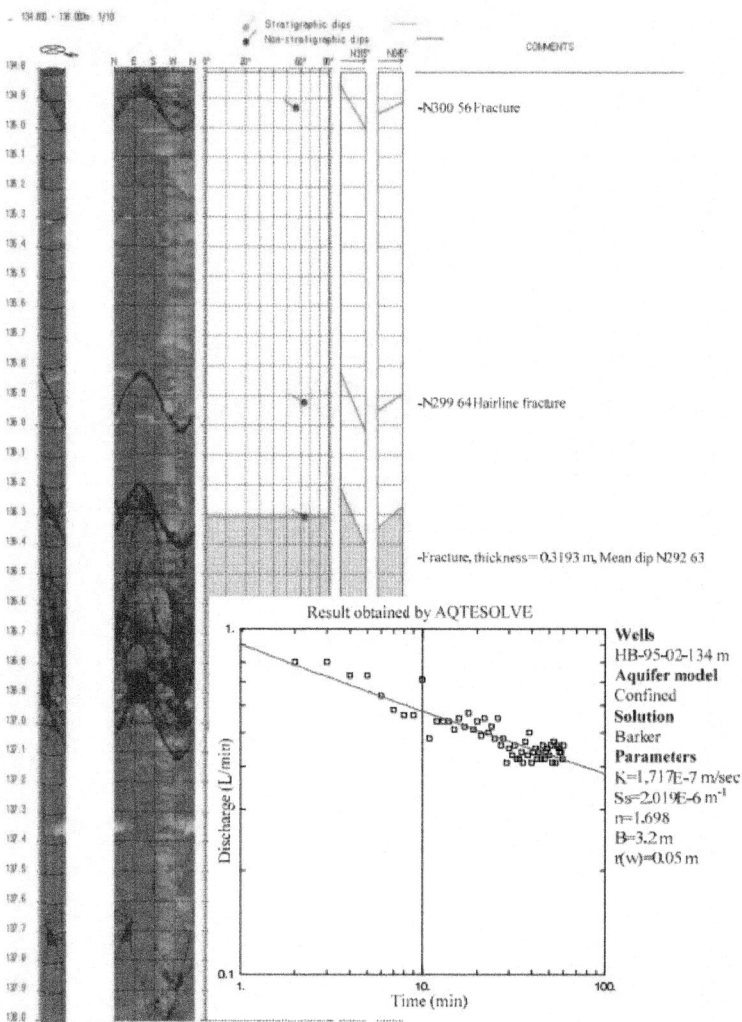

Figure 5. (b) Evaluation of hydraulic parameters using AQTESOLVE (right lower figure) and BHTV images at pack-off zones 134.8 m to 138 m in Borehole HB-95-02.

A total of 26 hydraulic tests were designed to determine hydraulic conductivity in the boreholes. Data collected during hydraulic tests can be analyzed by analytical methods. Water pressure and discharge rate measurements with time for each hydraulic test were collected in this study. The data analysis was performed using a professional version of the AQTESOLVE test analysis software, which enables both virtual and automatic type curve matching (Duffield, 2004). The quantitative evaluation of hydraulic parameters was carried out as an iterative process of the best-fit theoretical response curves based on the measured data of the hydraulic test.

EMPIRICAL MODEL OF ROCK MASS HYDRAULIC CONDUCTIVITY

Prior to describing the empirical model of rock mass hydraulic conductivity, an attempt to find the decrease in permeability with depth was conducted. Figure 6 demonstrates that the testing data of HB-94-01 shows the tendency that the hydraulic conductivity decreases with depth. The form of the regression equation is close to the result obtained by Black, 1987. The coefficient of determination of the regression equation is 0.633. However, the testing data from HB-95-01 and HB-95-02 are very scattered. No relationship can be found between hydraulic conductivity and depth. Accordingly, potential factors, including rock quality designation (RQD), depth index (DI), gouge content designation (GCD), and lithology permeability index (LPI), that may affect the degree of permeability should be considered. The rating approach for each factor that represents the magnitude of permeability is also described as following.

Rock Quality Designation

To assess the influence of the fracture characteristic on permeability, the rock quality designation (RQD) index (Deere et al., 1967), can be adopted. The RQD index was introduced over 40 years ago as an indicator of rock mass conditions. The RQD value is defined as the cumulative length of core pieces longer than 100 mm in a run (RS) divided by the total length of the core run (RT) and can be obtained from the following equation.

$$RQD = \frac{\sum \text{Length of Intact and Sound Core Pieces} > 100 \text{ mm}}{\text{Total Length of Core Run, mm}} \times 100\%$$

$$= \frac{R_S}{R_T} \times 100\% \tag{1}$$

In this study, a core run for calculating a RQD value is herein defined as a selected zone of a hydraulic test. Eq. 1. may be utilized to identify rock mass permeability.

Depth Index

The decrease in permeability with depth in fractured rocks is usually attributed to reduction in fracture aperture and fracture spacing. The reduction is due to the effect of geostatic stresses, and thereby the permeability of fractured rocks will be reduced. The depth may be considered as a factor in evaluating rock mass permeability. To assess the influence of the depth on permeability, a Depth Index, namely DI, was defined as the following equation.

$$DI = 1 - \frac{L_c}{L_T}$$

(2)

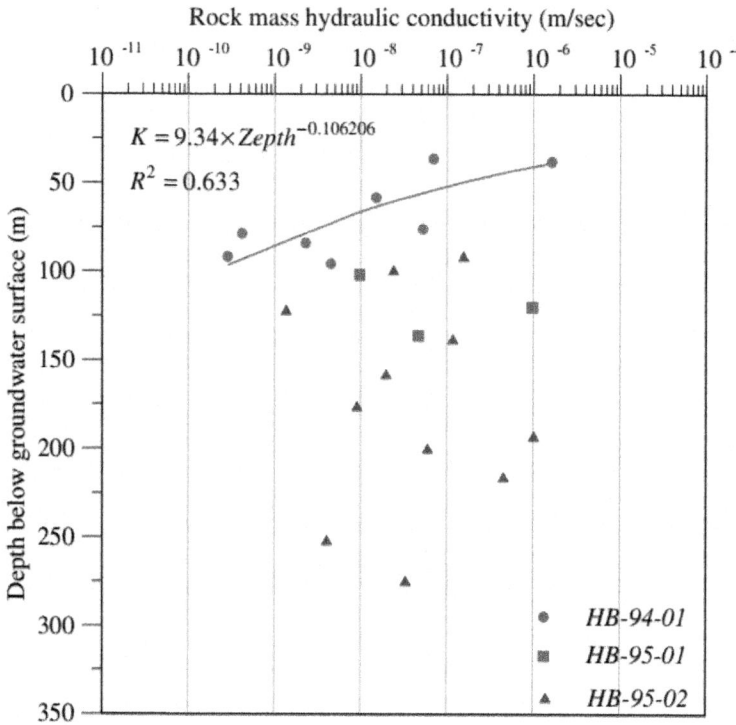

Figure 6. Relationship between hydraulic conductivity and depth.

in which LT is the total length of a borehole; LC is a depth which is located at the middle of a double packer test interval in the borehole. The value of DI is always greater than zero and less than one. The greater the DI value, the higher the permeability.

Gouge Content Designation

The RQD value may decrease by an increase of fractures in a core run. If the fractures contain infillings such as gouges, permeability of the fractures will reduce. To assess the influence of the gouge materials on permeability, a Gouge Content Designation (GCD) index was defined as the following equation.

$$GCD = \frac{R_G}{R_f - R_s},$$

(3)

in which RG is the total length of gouge content. The value of GCD is always greater than zero and less than one. The greater GCD value stands for the more gouge content in a core run, and thereby it will reduce the permeability.

Lithology Permeability Index

Lithology is the individual character of a rock in terms of mineral composition, grain size, texture, color, and so forth. For an intact rock, the magnitude of permeability depends largely on the individual character of the rock. It may be affected by the average size of the pores, which in turn is related to the distribution of particle sizes and particle shape. In sedimentary formations grain-size characteristics are most important because coarsegrained and well-sorted material will have high permeability as compared to fine-grained sediments like silt and clay. Thus, the lithology may be regarded as a factor in evaluating rock mass permeability. To assess the influence of lithology on permeability, a Lithology Permeability Index (LPI) was defined as Table 2.

Table 2. Description and ratings for lithology permeability index.

Lithology	Hydraulic conductivity (m/s)				Range of rating	Suggested Rating
	Reference[1]	Reference[2]	Reference[3]	$K_{average}$		
Sandstone	10^{-6}~10^{-9}	10^{-7}~10^{-9}	10^{-7}~10^{-9}	$10^{-7.5}$	0.8-1.0	1.00
Silty Sandstone	—	—	—	—	0.9-1.0	0.95
Argillaceous Sandstone	—	—	—	—	0.8-0.9	0.85
S.S. interbedded with some Sh.	—	—	—	—	0.7-0.8	0.75
Alternations of S.S & Sh.	—	—	—	—	0.6-0.7	0.65
Sh. interbedded with some S.S.	—	—	—	—	0.5-0.7	0.60
Alternations of S.S & Mudstone	—	—	—	—	0.5-0.6	0.55
Dolomite	10^{-6}~$10^{-10.5}$	10^{-7}~$10^{-10.5}$	10^{-9}~10^{-10}	10^{-8}	0.6-0.8	0.70
Limestone	10^{-6}~$10^{-10.5}$	10^{-7}~10^{-9}	10^{-9}~10^{-10}	10^{-8}	0.6-0.8	0.70
Shale	10^{-10}~10^{-12}	10^{-10}~10^{-13}	—	$10^{-10.5}$	0.4-0.6	0.50
Sandy Shale	—	—	—	—	0.5-0.6	0.60
Siltstone	10^{-10}~10^{-12}	—	—	10^{-11}	0.2-0.4	0.30
Sandy Siltstone	—	—	—	—	0.3-0.4	0.40
Argillaceous Siltstone	—	—	—	—	0.2-0.3	0.20
Claystone	—	10^{-9}~10^{-13}	—	10^{-11}	0.2-0.4	0.30
Mudstone	—	—	—	—	0.2-0.4	0.20
Sandy Mudstone	—	—	—	—	0.3-0.4	0.40
Silty Mudstone	—	—	—	—	0.2-0.3	0.30
Granite	—	—	10^{-11}~10^{-12}	$10^{-11.5}$	0.1-0.2	0.15
Basalt	10^{-6}~$10^{-10.5}$	10^{-10}~10^{-13}	—	$10^{-11.5}$	0.1-0.2	0.15

[1]B.B.S. Singhal & R.P. Gupta (1999) ; [2]Karlheinz Spitz & Joanna Moreno (1996) ; [3]Bear(1972)

Rock Mass Permeability Index

As stated above, the rock mass permeability may be dependent on the following four parameters: RQD, DI, GCD, and LPI. However, the permeability is not simply affected by only one factor. It may account for the synthetic effect from the four parameters on permeability. Accordingly, Rock mass permeability index, called the HC index, was proposed herein.

$$HC = (1-RQD)(DI)(1-GCD)(LPI)., \qquad (4)$$

The value of each parenthesis at the right hand side of Eq. 4. is always greater than zero and less than one depending on the values assigned to the four parameters. The greater the value of each parenthesis, the higher the permeability. Thus, the model performs a numerical assessment of rock mass permeability using the four parameters. Since it is rare to encounter the condition that RQD is 100% in highly disturbed clastic sedimentary rocks in Taiwan, the term of (1-RQD) is usually greater than zero. However, it should be noted that if (1-RQD) is zero, the value of 0.01 in the term of (1-RQD) is suggested to avoid the HC value to be zero. Currently, the study took the same weight for each factor in Eq. (4). In addition, Eq. (4) is limited in sedimentary rocks only and is only applied to vertical boreholes at present. With more testing data, a further study can be considered to assign a different weight for each factor to give a better correlation between the hydraulic conductivity and HC.

The Empirical HC Model

Regression analysis was performed to estimate the dependence of HC on hydraulic conductivity. A total of 22 hydraulic test data were applied to the study. HC-values for the hydraulic tests can be computed from borehole image data and rock core data, in which the values of RQD and GCD at each test interval can be calculated from borehole image data and rock core data with Eqs. 1. and 3., respectively. The value of DI can be calculated using Eq. 2. The value of LPI for each test zone can be obtained from rock core data and Table 3. Table 3 shows the calculated results for the HC model based on the verified data. The regression results indicated that a power law relationship exists between the hydraulic conductivity and HC with a coefficient of determination of 0.866 as shown in Fig. 7. The empirical HC model is obtained as shown in Eq. 5.

$$K = 2.93 \times 10^{-6} \times (HC)^{1.380}, R^2 = 0.866 \qquad (5)$$

If only HB-94-01 testing data were adopted, a better correlation with the coefficient of determination of 0.905 can be obtained as shown in Eq. 6.

$$K = 2.31 \times 10^{-6} \times (HC)^{1.342}, R^2 = 0.905 \tag{6}$$

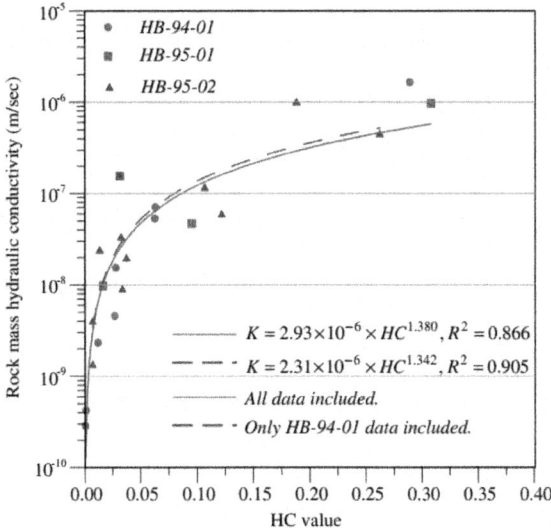

Figure 7. Relationship between hydraulic conductivity and HC-values.

Table 3. The calculated results for HC-system based on 22 hydraulic test data.

Boreholes	Test intervals (m)	1-RQD	DI	1-GCD	LPI	HC	K (m/s)
HB-94-01	34.7-36.3	0.094	0.677	1.000	1.000	0.0635	7.06E-08
	36.4-38.0	0.438	0.662	1.000	1.000	0.2895	1.64E-06
	56.7-58.3	0.063	0.477	1.000	0.950	0.0283	1.53E-08
	74.6-76.2	0.500	0.315	1.000	0.400	0.0629	5.3E-08
	77.2-78.8	0.010	0.291	1.000	0.400	0.0012	4.22E-10
	82.6-84.2	0.125	0.242	1.000	0.400	0.0121	2.31E-09
	90.2-91.8	0.010	0.173	1.000	0.400	0.0007	2.86E-10
	94.2-95.8	0.500	0.136	1.000	0.400	0.0273	4.53E-09
HB-95-01	99.0-101.9	0.345	0.598	0.200	0.400	0.0165	9.8E-09
	117.2-120.1	0.690	0.526	1.000	0.850	0.3081	9.76E-07
	133.2-136.1	0.724	0.461	0.286	1.000	0.0954	4.68E-08
HB-95-02	88.6-91.4	0.071	0.743	1.000	0.600	0.0318	1.56E-07
	96.0-99.2	0.031	0.721	1.000	0.600	0.0135	2.42E-08
	118.5-121.7	0.219	0.657	0.071	0.700	0.0072	1.36E-09
	134.8-138.0	0.344	0.610	0.727	0.700	0.1068	1.17E-07
	154.8-158.0	0.938	0.553	0.103	0.700	0.0376	1.99E-08
	173.0-176.2	0.938	0.501	0.103	0.700	0.0340	9.08E-09
	189.8-193.0	0.594	0.453	1.000	0.700	0.1883	1.01E-06
	196.6-199.8	0.563	0.434	0.500	1.000	0.1220	6.00E-08
	213.2-216.0	0.679	0.387	1.000	1.000	0.2625	4.54E-07
	249.0-251.8	0.393	0.285	0.091	0.700	0.0071	4.03E-09
	272.0-274.8	0.214	0.219	1.000	0.700	0.0328	3.36E-08

It should be noted that the values of (1-GCD) in HB-94-01 borehole are all equal to 1. The results of Eq. 6 demonstrate that the empirical HC model may also be more accurate for the estimation of the rock mass hydraulic conductivity if the fractures do not contain infillings. There are a few limitations that need to be noted for the use of Eq. 5. The data used to develop the equation are limited in number and in the lithologies represented. From the definition of DI, DI cannot be determined for inclined boreholes because the data collected were from vertical boreholes.

Model Verification

In order to further verify the feasibility of the proposed empirical HC model, the model verification is conducted. Another borehole data with the drilling depth of 120 m is adopted to verify the empirical HC model. The principal lithologic units of the borehole, namely CH- 04, are mainly sandstone, shale, and sandstone with some thin shale. The depth from 24.5 m to 26.6 m, 32.5 m to 34.1 m, 65.7 m to 67.8 m, and 77.8 m to 79.9 m were sealed by double packers for conducting the hydraulic tests. The quantitative evaluation of hydraulic parameters was then performed using AQTESOLVE which uses an iterative process of the best-fit theoretical response curves based on the measured data of the hydraulic test. Figure 8 shows that the comparison of the rock mass hydraulic conductivity obtained by in-situ test and that from the estimation of the empirical HC model. Very good correlation can be found (Fig. 8). This verification example demonstrates that the empirical HC model is able to determine the rock mass hydraulic conductivity for different sites in which the lithologic conditions are similar.

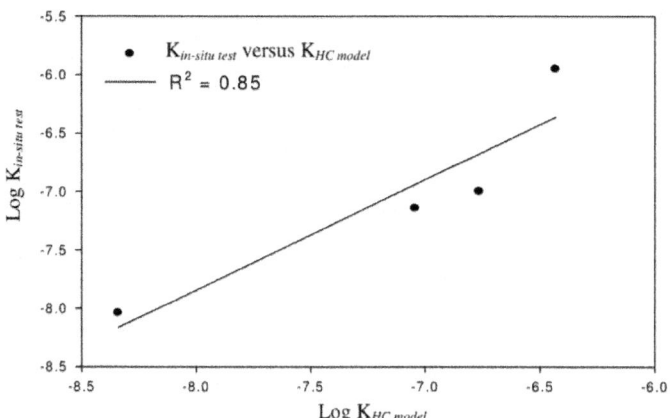

Figure 8. Correlation between Kin-situ and KHC model.

Table 4. Four hydraulic test data for the model verification (Borehole CH-04).

Test intervals (m)	RQD(%)	DI	1-GCD	LPI	HC	$K_{HC\ model}$	$K_{in\text{-}situ}$
24.5~26.6	81.0	0.787	0.952	0.55	0.0785	9.06E-08	7.14E-08
32.5~34.1	43.8	0.723	0.975	0.55	0.2179	3.69E-07	1.11E-06
65.7~67.8	47.6	0.444	0.976	0.55	0.1248	1.71E-07	9.95E-08
77.8~79.9	95.2	0.343	1.000	0.55	0.0090	4.59E-09	9.09E-09

$K_{HC\ model}$ and $K_{in\text{-}situ}$ represent K obtained by Eq. 5 and the in-situ hydraulic test, respectively.

CONCLUSIONS

The estimation of rock mass hydraulic conductivity of highly disturbed clastic sedimentary rocks in Taiwan was performed using the data of BHTV and double packer hydraulic tests. The field results indicated that the rock mass in the study area has the conductivity between the order 10-10 and 10-6 m/s at the depth between 34 m and 275 m below ground surface. The results demonstrate that the rock mass hydraulic conductivity of highly disturbed clastic sedimentary rocks in Taiwan mainly depends on the following four parameters: RQD, DI, GCD, and LPI.

This paper proposes an empirical HC model for estimating rock mass hydraulic using data collected for highly disturbed clastic sedimentary rocks in Taiwan. The HC-value can be calculated from borehole image data and rock core data. To verify rationality of the proposed HC model, the study collected data from the results of two hydrogeological investigation programs in three boreholes to determine a relationship between hydraulic conductivity and HC. Besides, good correlation is found from the model verification which demonstrates that the empirical HC model is able to determine the rock mass hydraulic conductivity for different sites in which the lithologic conditions are similar. The regression results indicated that the relationship of a power law exists between the two variables with a coefficient of determination of 0.866. The empirical HC model may provide a useful tool to predict hydraulic conductivity of fractured rocks based on measured HCvalues. By using this model, hydraulic conductivity data in a given site can be directly acquired, which removes the cost on hydraulic tests. For in-situ aquifer tests, the empirical HC model is valuable for preliminary assessment of the degree of permeability in a packedoff interval of a borehole.

REFERENCES

1.   Bear, J. (1972). Dynamics of Fluids in Porous Materials, American Elsevier.

2. Black, J. H. (1987). Flow and flow mechanisms in crystalline rock, in Fluid Flow in Sedimentary Basins and Aquifers. Geol. Soc. Special Publication No. 34, 186-200.

3. Burgess, A. (1977). Groundwater Movements Around a Repository — Regional Groundwater Analysis. Kaernbraenslesaekerhet, Stockholm, Sweden, 116.

4. Carlson, A. & Olsson, T. (1977). Hydraulic properties of Swedish crystalline rocks-hydraulic conductivity and its relation to depth. Bulletin of the Geological Institute, University of Uppsala 7, 71-84.

5. Deere, D. U.; Hendron, A. J.; Patton, F. D.& Cording, E. J. (1967). Design of surface and near surface construction in rock. Proceedings of 8th U.S. Symposium Rock Mechanics, AIME, New York 237-302.

6. Duffield, G. M. (2004). AQTESOLVE version 4 user's guide, Developer of AQTESOLV HydroDOLVE, Inc., Reston, VA, USA.

7. National Research Council. (1996). Rock fractures and fluid flow: contemporary understanding and applications. National Academy Press, Washington D. C., USA.

8. Singhal, B. B. S. & Gupta, R. P.(1999). Applied hydrogeology of fractured rocks. Kluwer Academic Publishers, The Netherlands, 400.

9. Spitz, K. & Morena, J. (1996). A Practical Guide to Groundwater and Solute Transport Modeling, Wiley.

10. Sinotech Engineering Consultants, LTD. (2006). Tseng-Wen transbasin diversion tunnel project-supplemental geology investigation, Southern Water Resources Office, Water Resources Agency, Ministry of Economic Affairs, Taiwan (in Chinese).

11. Snow, D. T. (1969). Anisotropic permeability of fractured media. Water Resources Research, 5(6), 1273-1289.

12. Wei, Z.Q., Egger, P., Descoeudres, F. (1995). Permeability predictions for jointed rock masses. International Journal of Rock Mechanics, Mineral Science and Geomechanics 32, 251-261.

13. Williams, J. H. & Johnson, C. D. (2004). Acoustic and optical borehole-wall imaging for fractured-rock aquifer studies. Journal of Applied Geophysics, 55(1–2): 151–159.

14. Zhao, J. (1998). Rock mass hydraulic conductivity of the Bukit Timah granite, Singapor. Engineering Geology, V 50, 211-216.

Chapter 8

GENERAL HYDRAULIC GEOMETRY

Levent Yilmaz[1]
[1] Technical University of Istanbul, Civil Engineering Department, Hydraulics Division, Maslak, Istanbul, Turkey

ABSTRACT

Employing bed load formulae hydraulic geometry relations were derived for stream width, meander wave length, and bed slope. The relations are in terms of friction factor, bed load discharge, bed load diameter, and water discharge. The bed load formulae are those of Engelund and Hansen (1966) [1], Einstein (1950) [2], Shields (1936) [3], and Meyer-Peter and Muller (1948) [4].

INTRODUCTION

The water and sediment discharge of a river are primarily determined by the hydrology, geology, and topography of the drainage area. According to the influx water and sediment, the river creates its own geometry, i.e., slope, depth, width, and meandering pattern. Since the slope and meander pattern do not respond quickly enough to follow seasonal variations of discharge, it is natural to invoke some measure of dominant discharge values for these variables (Engelund and Hansen, 1966 [1] ; Hansen, 1967 [5]; Kennedy and Alam, 1967 [6]).

In a given river, the water and sediment discharges normally increase in the downstream direction, and so do the depth and the width of the stream. The slope and the grain size usually decrease gradually from the source to estuary. According to Leviavsky (1955) [7], the grain size decreases approximately exponentially in the downstream direction. These observations point towards the existence of relations for the depth, width, and slope as functions of the water and sediment discharge. To that end, a number of "regime" relations have been suggested using it. An account of such relations has been given by Blench (1957[8], 1966[9]). Accordingly,

$$B \sim Q^{1/2} \tag{1}$$

$$D \sim Q^{1/3} \tag{2}$$

$$S \sim Q^{-1/6} \tag{3}$$

where B = the width, D = the depth, S = the slope, and Q = the water discharge. An empirical relation expressing the meander " wave-length" has been suggested by Inglis (1947) [10] as:

$$L \sim Q^{1/2} \tag{4}$$

Equations (1) and (4) describe a direct proportionality between the width of the stream and the meander length. Using the data from a large number of rivers in U.S.A and Indian as well as from several small scale model tests, Leopold and Wolman (1957) [11]derived the following popular relation:

$$L = 10B \tag{5}$$

Engelund and Hansen (1967)[12] derived equation (5) using the similarity principle. Due to the complex mechanics of the bed load and water transport, a number of variations of the above formulae have been proposed in the literature.

The objective of this paper is to derive hydraulic geometry relations using bed load formulae and compare them.

DERIVATION OF HYDRAULIC GEOMETRY RELATIONS

Engelund and Hansen (1966) [1] Bed Load Formula

The Darcy-Weisbach relation for the energy slope can be expressed as:

$$S = f \frac{V^2}{2g} \frac{1}{D} \tag{6}$$

where S = energy gradient (slope), f = friction factor, V= mean velocity, g = acceleration due to gravity, D = mean flow depth. Engelund and Hansen (1967) [12] expressed f as

$$f = 0.1\theta^{5/2}/\Phi \tag{7}$$

where θ = the dimensionless form of the bed shear stress τ_0, and Φ= non-dimensional sediment transport rate, and d = mean fall diameter.

The non-dimensional transport rate is expressed as

$$\Phi = \frac{q_T}{\sqrt{(s-1)gd^3}}$$

(8)

where $q_T = Q_T/B =$ sediment discharge per unit width, $Q_T =$ total sediment discharge ($= Q_B + Q_S$), $Q_B =$ bed load discharge, $Q_S =$ suspended load discharge, $s = \gamma s/\gamma =$ relative density of sediment grains, $\gamma =$ specific gravity of water, $\gamma_s =$ specific gravity of sediment grains, B = flow width,. Substitution for $q_T = Q_T/B$ in equation (8) yields

$$\Phi = \frac{Q_T}{B\sqrt{(s-1)gd^3}}$$

(9)

The bed shear shear stress τ_0, can be expressed in dimensionless form, θ, as (Shields, 1936)[3]:

$$\theta = \frac{DS}{(s-1)d}$$

(10)

Substituting equations (7) – (10) into the Darcy-Weisbach equation (6) one gets

$$S = 0.1 \left[\frac{DS}{(s-1)d} \right]^{5/2} \frac{1}{\Phi} (V)^2 \frac{1}{2g} \frac{1}{D}$$

(11)

Substitution of equation (10) for Φ, the non-dimensional transport rate and the continuity equation V=Q/BD in equation (11) yields

$$S = 0.1 \left[\frac{DS}{(s-1)d} \right]^{5/2} \frac{B\sqrt{(s-1)gd^3}}{Q_T} (Q/BD)^2 \frac{1}{2g} \frac{1}{D}$$

(12)

Recalling the definition of s and putting the values of s = 2.65 and g = 9.81 in equation (12), one obtains:

$$S = 0.1 \left[\frac{DS}{(1.65)d} \right]^{5/2} \frac{B\sqrt{(1.65)gd^3}}{Q_T} \left(\frac{Q}{BD} \right)^2 \frac{1}{19.62} \frac{1}{D}$$

(13)

A little rearrangement of equation (13) yields

$$\frac{S}{S^{5/2}} = \frac{0.005}{(1.65)^{5/2}} d^{-5/2} D^{5/2} Q^2 / Q_T [(1.65)(9.81)]^{1/2} d^{3/2} BD^{-3}$$

(14)

Equation (14) can be simplified as

$$S = \left[0.20\ d^{-1} BD^{1/2} \left(Q^2/Q_T \right) \right]^{-2/3}$$

(15)

The mean depth can be expressed from equation (15) as

$$D = \left[5S^{-3/2}\,Bd\,(Q_T/Q^2)\right]^2 \tag{16}$$

The water surface width B can be expressed from equation (16) as

$$B = 0.20\,D^{1/2}S^{3/2}d\,(Q^2/Q_T) \tag{17}$$

or using L = 10B, one obtains from equation (17):

$$L = 2.0\,D^{1/2}S^{3/2}d\,(Q^2/Q_T) \tag{18}$$

Shields' (1936) [3] Bed Load Formula:

On the basis of his experimental results, Shields (1936) [3] proposed a dimensionally homogeneous transport function:

$$Q_B\gamma_s/Q\gamma S = 10(\tau_0 - \tau_c)/(\gamma_s - \gamma)d_m \tag{19}$$

where Q_B = bed load transport rate (tons/hour), d_m = the effective diameter of sediment (mm), γ_s = the specific weight of sediment (T/m³), γ = the specific weight of water (T/m³), τ_0 = the shear stress (T/m²); τ_c = critical shear stress (T/m²), and S = the energy slope.

Shields (1936) [3] hypothesized that the rate of transport was a function of the dimensionless resistance coefficient:

$$\tau_0\left[(\gamma_S - \gamma)d_m\right]^{-1} \tag{20}$$

the critical value controlling the incipient motion of the bed load. The computation of the shear stress uses the hydraulic radius R and the bed slope S:

$$\tau_0 = \gamma RS \tag{21}$$

whereas the critical shear stress is obtained from Straub's graph (1935) [3] for various sediment sizes. The bed load transport rate obtained from the Shields formula is the mass

of the solid particles. This rate value should be multiplied by a factor 1.60, in accordance with the density of sand, to obtain the volumetric value.

Recalling that

$$\tau_0 - \tau_c = \tau = \gamma DS \tag{22}$$

Substituting of equation (22) in equation (19) gives

$$\frac{\gamma_s Q_B}{\gamma QS} = \frac{10\gamma DS}{(\gamma_s - \gamma)d_m} = \frac{10DS}{\left(\frac{\gamma_s}{\gamma} - 1\right)d_m} \tag{23}$$

where $s = \dfrac{\gamma_s}{\gamma}$. This can be simplified as

$$\frac{\gamma_s Q_B}{\gamma Q S^2} = \frac{10D}{\left(\frac{\gamma_s}{\gamma} - 1\right) d_m}$$

(24)

Substituting $Q_B = q_B B$ and equation (6) into equation (24), one obtains

$$\frac{q_B B D 4 g^2 B^4 D^4}{Q f^2 Q^4} = \frac{10D}{\frac{\gamma_s}{\gamma}\left(\frac{\gamma_s}{\gamma} - 1\right) d_m}$$

(25)

Using the continuity equation V=Q/A, one gets from equation (25)

$$Q_B = \frac{10 Q^5 f^2}{\frac{\gamma_s}{\gamma}\left(\frac{\gamma_s}{\gamma} - 1\right) d_m B^4 D^5 4 g^2}$$

(26)

D is expressed from equation (26) as

$$D = 0.3587 \frac{Q}{Q_B^{0.20}} f^{0.4} d_m^{-0.2} B^{-0.8}$$

(27)

Using L=10B, one obtains from equation (27)

$$L = 2.776 \frac{Q^{1.25}}{Q_B^{0.25}} f^{0.5} d_m^{-0.25} D^{-1.25}$$

(28)

From equation (28) the width B is computed as

$$B = (0.2776) \frac{Q^{1.25}}{Q_B^{0.25}} f^{0.5} d_m^{-0.25} D^{-1.25}$$

(29)

S is derived from equation (23)

$$S = 0.6612 \left(\frac{Q_B}{Q}\right)^{0.5} d^{0.5} D^{-0.5}$$

(30)

Einstein's (1950) [2] Bed Load Formula

The bedload formula due to Einstein (1950) [2] can be expressed as

$$q_{sbi} = P_i \Phi_* \gamma_s / \left\{ [\gamma/\gamma_s - \gamma]/g d_{si}^3 \right\}^{\frac{1}{2}}$$

(31)

where d_{si} = the grain diameter for which si per cent is finer, q_{sbi} = the intensity of bed load movement of size class i, Φ_* = the bed load intensity, γ = the unit weight of water; γ_s = the unit weight of sediment particles, g = the gravitational acceleration, P_i = the particle availability parameter~ G_{di}/G_d (bed surface gradation),G_d = the total weight of sediment in the bed surface layer, and G_{di} = the weight of the i th size class in the bed surface layer. The bed surface layer in this equation is a zone near the bed surface called the "active layer". The surface elevation is changed as sediment is deposited into or scoured out of this layer.

Setting the particle availability parameter, P_1, equal to 1. One obtains from equation (31)

$$\frac{Q_B}{B} = \frac{\Phi_* \gamma_s}{\left[(\gamma/\gamma_s - \gamma)/gd_{si}^3\right]^{1/2}} \qquad q_s = \frac{Q_B}{B} \qquad B = \text{water surface width}$$

$$\tag{32}$$

where $q_s = P_i \Sigma q_{sbi}$ and $q_{sbi} = $ the intensity of bed load movement of size class i.

The water surface width can finally be expressed from the equation (32) as

$$B = Q_B \Phi_*^{-1} \gamma_s^{-1} \left[\left(\gamma^2/(s-1)\gamma_s\right)/gd_{si}^3\right]^{1/2} \tag{33}$$

Using L=10B, one obtains from the equation (36) as

$$L = 10 Q_B \Phi_*^{-1} \gamma_s^{-1} \left[\left(\gamma^2/(s-1)\gamma_s\right)/gd_{si}^3\right]^{1/2} \tag{34}$$

The Darcy-Weisbach relation, equation (6) can be written as

$$S == f \frac{Q^2}{B^2 D^3} \frac{1}{2g} \tag{35}$$

Substituting the water surface width B from the equation (32) into the equation (35) and using the continuity equation Q=V BD,

D is expressed from Equation (35) as

$$D = 0.2529 f^{0.571} Q^{0.571} B^{-0.571} V^{0.28571} \tag{36}$$

The mean velocity is derived from equation (36) as

$$V = 122.93 f^{-1.998} Q^{-1.998} B^{1.998} D^{3.5} \tag{37}$$

and the slope S is derived from the equation (38) as

$$S = f Q^2 B^2 0.0509 D^{-3} \tag{38}$$

Bed Load Formula of Meyer-Peter and Muller (MPM) Model (1948) [4]

This formula was derived from experiments using a laboratory flume with a maximum width of 2 m, very different from the conditions encountered in large channels. The formula depends primarily on the grain diameter and water discharge. They derived a formula for bed load discharge with the aim to develop a more practical formula. Bogardi (1978) indicated several difficulties that are encountered in application of this formula.

According to the Meyer-Peter and Muller(1948) [4] (MPM) Model:

$$\gamma DS = 0.047(\gamma_s - \gamma)d_s + 0.25\rho^{1/3}q_b^{1/3} \tag{39}$$

where γ = specific gravity of water, γ_s = specific gravity of sediment grains, D= mean depth, S= energy gradient (slope), d_s= mean sediment diameter, ρ = density of water, q_b= sediment discharge per unit width.

From equation (39) the slope S is computed as

$$S = (\gamma D)^{-1}\left[0.047(\gamma_s - \gamma)d_s + 0.25\rho^{1/3}q_b^{1/3}\right]$$

(40)

Putting equation (6) into the equation (40)

$$\gamma f \frac{Q^2}{B^2 D^2} \frac{1}{2g} = 0.047(\gamma_s - \gamma)d_s + 0.25\rho^{1/3}q_b^{1/3}$$

(41)

Equation (41) is rearranged for computing the width B as

$$B^{-2} = 19.62D^2\left[0.047(\gamma_s - \gamma)d_s + 0.25\rho^{1/3}q_b^{1/3}\right]f^{-1}\gamma^{-1}Q^{-2}$$

(42)

From the equation (42), is computed the width B is computed

$$B = 0.226D^{-1}\left[0.047(\gamma_s - \gamma)d_s + 0.25\rho^{1/3}q_b^{1/3}\right]^{1/2}f^{1/2}\gamma^{1/2}Q^{-1}$$

(43)

Using L=10B, one obtains the wave length L as

$$L = 2.26D^{-1}\left[0.047(\gamma_s - \gamma)d_s + 0.25\rho^{1/3}q_b^{1/3}\right]^{1/2}f^{1/2}\gamma^{1/2}Q^{-1}$$

(44)

The water depth D is derived from equation (42) as

$$D = 0.226B^{-1}\left[0.047(\gamma_s - \gamma)d_s + 0.25\rho^{1/3}q_b^{1/3}\right]^{1/2}f^{1/2}\gamma^{1/2}Q^{-1}$$

(45)

DISCUSSION

The hydraulic geometry relationships, B, D, S, L computed with DuBoys (1879) [13] sediment transport formula (Huang and Nanson, 2000) [14] were similar, those computed with Einstein's (1950) [2] model and Shields' (1936) [3] model, but MPM (1948) [4] model have differences by using the specific gravity of water and sediment grains, and density of water. Although there are four dependent variables (width, depth, velocity and slope) with only three basic flow relations of continuity, resistance and sediment transport. This study finds that the long-recognized problem of no closure can be solved directly in terms of the analytical approach advocated here for understanding the self-adjusting mechanism of alluvial channels.

With stable canal flow relations (Lacey's flow resistance relation and DuBoys' (1879) [13] sediment transport formula) and rectangular sections, introducing a channel form factor (width/depth ratio) as a dependent variable identifies an optimum condition for sediment transport by adjusting width/

depth ratio for a given flow discharge, channel slope and sediment size (Huang and Nanson, 2000) [14].

Theoretically derived channel geometry relations are highly consistent with their counterparts obtained from " Darcy-Weisbach relation", except that ' threshold theory' provides a larger value of friction factor coefficient. This may be due to the use of rectangular cross-sections in our analytical study. Furthermore, the maximum friction value is greater in natural rivers using relationships not so dependent on canal data.

A comparison of the averaged channel geometry relations with downstream hydraulic geometry relations developed by Huang and co-workers (Huang and Warner, 1995 [15]; Huang and Nanson, 1995 [16], 1998 [17]) and by Julien and Wargadalam (1995) [18] based on numerous sets of field observations, reveals high level of consistency. When sediment concentration varies in a limited range, the averaged relationships are very similar to empirical regime formulations ('regime theory') (Huang and Nanson, 2000) [4].

Table I presents a comparison of equations with downstream hydraulic geometry relations obtained by Huang and co-workers (2000) and by Julien and Wargadalam (1995) [18], based on numerous sets of field observations.

Table 1. Downstream hydraulic geometry relations defined as the functions of flow discharge and channel slope.

Huang and co-workers' model (1995) [16] $B \propto Q^{0.501}S^{-0.156}$ $D \propto Q^{0.299}S^{-0.206}$ $V \propto Q^{0.200}S^{0.362}$	Julien and Wargadalam's model(1995)[18] $B \propto Q^{0.4-0.5}S^{-(0.2-0.25)}$ $D \propto Q^{0.4-0.25}S^{-(0..2-0.125)}$ $V \propto Q^{0.2-0.25}S^{0.4-0.375}$
Huang and Nanson (2000) [14] $B \propto Q^{0.478}S^{0.076}$ $D \propto Q^{0.289}S^{-0.350}$ $V \propto Q^{0.233}S^{0.274}$	Engelund and Hansen (1967) model [12] $B \propto Q^2(Q^{-4}S^{-3})^{0.5}S^{1.5}$ $D \propto Q^{-4}S^{-3}$
Shields (1936) model [3] $B \propto Q^{1.25}f^{0.5}D^{-1.25}d_m^{-0.25}$ $D \propto (Q/Q_B^{0.20})f^{0.4}d_m^{-0.2}B^{-0.8}$	
Einstein's (1950) model [2] $B \propto \Phi_*^{-1}d_{si}^{1.5}$ $D \propto S^{-0.33}f^{0.33}Q^{0.66}(\Phi_*^{-1}d_{si}^{1.5})^{0.66}$	
MPM (1948) model [4] $B \propto Q^{-1}f^{0.5}D^{-1}$ $D \propto Q^{-1}f^{0.5}B^{-1}$	

To reflect how channel geometry adjusts within the range, this study is only able to present acceptably averaged relationships for channel geometry by assigning the four variables depth, width, slope and wave length.

Although as stated earlier τ_c and the other coefficients are determined only by sediment size d, their combined effects on natural channel geometry are much more complicated with both bank strength and channel roughness (or sediment size) being often particularly examined (Millar and Quick, 1993 [19]; Huang and Nanson, 1998 [17]). When the constant terms and the terms related to sediment size d are ignored, the equations of Einstein's (1950) [2] and Shields' (1936) [3] agree very closely with the widely observed empirical regime channel relationships, when sediment concentration (Q_s/Q) remains unchanged or varies within a limited range. This is consistent with the study by Simons and Albertson (1960) [20]. In an analysis of numerous observations that were collected from stable canals in different parts of the world, Simons and Albertson (1960) [20] identified that the relations closest to this equations occur only when sediment concentration varies in a limited range (less than 500 ppm).

In many circumstances, sediment discharge Q_s is unknown and consequently channel slope S is used as an alternative. The consistency of the theoretical results of equations (Engelund and Hansen, 1967 [12]; Shields [3], 1936; Einstein, 1950 [2]; MPM model, 1948 [4]) with the studies based on direct observations suggests strongly that most natural alluvial channels are able to adjust their channel form, so as to reach an optimum state. This must be a general principle for flow in rivers and canals that causes channels in very different environments to exhibit remarkably similar hydraulic geometry relations.

The physical relationships of flow continuity, flow resistance and sediment transport determine the degree of channel adjustment and consequently illustrate a condition of maximum sediment transporting capacity, subject to the conditions of flow discharge, channel slope and sediment size.

Mathematically, it can be defined a minimum slope subject to the conditions of flow discharge, sediment discharge and sediment size; or a minimum flow discharge subject to the conditions of sediment discharge, channel slope and sediment size; or an optimum sediment size subject to the conditions of flow and sediment discharges, and channel slope. This formulation provides the maximum sediment transporting capacity per unit of approximate total stream power (actual total stream power is γQS where γ is the specific weight of water). This concept includes the following specific optimum conditions that have long been hypothesized and applied for practical problem solving:

- for fixed S and Q, Q_s = a maximum as proposed by Pickup (1976) [21], Kirkby (1977) [22] and White et al. (1982) [23];

- for fixed Q_s and Q, S= a minimum as proposed by Chang (1979 a [24], b [25], 1988 [26]);

- for fixed Q_s, $S^{7/11}Q^{8/11}$= a minimum or $SQ^{1.142}$= a minimum, close to γQS= a minimum as proposed by Chang (1980 a [27], b [28], 1986 [29], 1988 [26]).

Hence, the use of the analytical approaches proposed by Pickup (1976) [21], Kirkby (1977) [22] and White et al. (1982) [23] with the maksimum sediment transport capacity, and by Chang (1979 a [24], b [25], 1980 a [27], b [28], 1988 [26]) based on the minimum stream power hypothesis should produce consistent stable channel geometry relations with the theoretical approach advocated in this study.

This study shows that the optimum condition for sediment transport with regard to downstream hydraulic geometry relations for a given flow discharge, channel slope and sediment size results from the general condition of maksimum flow efficiency according to Huang and Nanson, 2000 [14], defined as the maximum sediment transporting capacity per unit available stream power. Maksimum flow efficiency is an internal optimum condition and includes the conditions of maximum sediment transporting capacity and minimum stream power as proposed by Pickup (1976) [21], Kirkby (1977) [22] and White et al. (1982) [23], and by Chang (1979 a [24], b [25], 1980 a [27], b [28], 1986 [29], 1988 [26]) respectively.

Despite criticism of the use of extremal hypotheses (Griffiths, 1984)[30], this study offers strong support for the use of bed-load concepts of different authors for hydraulic-geometry relation derivations and for understanding natural channel-form adjustment.

Finally, this study indicates that the general principle of sediment transport approaches in the variational theory of mechanics is able to provide a physical explanation for the existence of the optimum conditions of natural channel-form adjustment. However, these findings are of a preliminary nature and further detailed research is ongoing.

CONCLUSION

Strictly speaking, Engelund and Hansen's equation should be applied to flows with dune beds in accordance with the similarity principle. According to Yang(1987) [31], a river can adjust its roughness, geometry, profile and pattern through the processes of sediment transport. Qualitative descriptions of these dynamic adjustments of natural streams have been made mainly by geologists and geomorphologists. Empirical regime types of equation have been developed by engineers to solve design problems. Attempts have been made in recent years to explain these adjustments based on different extremal theories and hypotheses.

REFERENCES

1. Engelund, F. and E. Hansen, 1966: Investigations of Flow in Alluvial Streams. Acta Polytechnical Scandinavica. Civil Engineering and Building Construction Series No. 35, Copenhagen.

2. Einstein, H. A., 1950: The Bed-Load Function for Sediment Transportation in Open Channel Flows. Technical Bulletin No. 1026. U. S. Dept. of Agriculture.

3. Shields, I. A., 1936: Anwendung der Ahnlichkeitmechanik und der Turbulenzforschung auf die Geschiebebewegung. Mitt. Preuss. Ver. – Anst., Berlin, No. 26.

4. Meyer-Peter, E.,and R. Muller, 1948: Formulas for Bed-Load Transport. Proc. 3rd Meet. Int. Ass. Hydr. Res., Stockholm.

5. Hansen, E., 1967 : On the formation of Meanders as a Stability problem. Basic Research-Progress Report No. 13, Jan., Hydraulic Laboratory, Tech. Uni. of Denmark.

6. Kennedy, J. F., and A. M. Z. Alam, 1967: Discussion of "Hydraulic Resistance of Alluvial Streams" in Journ. Hydr. Div. Vol. 93, HY 1, January.

7. Leviavsky, S., 1955: An Introduction to Fluvial Hydraulics. Constable and Company Ltd., London.

8. Blench, T., 1957: Regime Behavious of Canals and Rivers. Butterworths Scientific Publications, London.

9. Blench, T., 1966 : Mobile-Bed Fluviology. Dept. Tech. Serv., Tech III. Div., Univ. of Alberta, Edmonto, Alberta, Canada.

10. Inglis, C. C., 1947: Meanders and their Bearing on River Training. Proc. Instn. Civ. Engrs., Maritime and Waterways Pap. No.7, January.

11. Leopold, L. B., and Wolman, M.G., 1957: River channel patterns: braided, meandering and straight. U.S. Geological Survey Profl. Paper, 282-B, pp. 39-85.

12. Engelund, F., and E. Hansen, 1967: Comparison Between Similarity Theory and Regime Formulae. Basic Research-Progress Report No. 13, January, Hydraulic Laboratory, Techn. Un. of Denmark.

13. Du Boys, P. 1879 : " Etudes du regime du Rhone et de I'action execee par les eaux sur un lit a fond de gravirs indefiniment affouillable", Annales

des Ponts et Chausees, Series 5, 18(49), part 2, pp. 141- 195

14. Huang, H. Q. and G.C.,Nanson, 2000 : Hydraulic Geometry and maximum flow efficiency as products of the principle of least action. Earth Surface Processes and Landforms, 25, pp. 1-16

15. Huang, H. Q. and R.F, Warner, 1995 : The multivariate controls of hydraulic geometry; a causal investigation in terms of boundary shear distribution. Earth Surfaces Processes and Landforms 20, pp. 115-130

16. Huang, H. Q. and G.C.,Nanson, 1995 : On a multivariate model of channel geometry. Proceedings of the XXVIth Congress of the International Association of Hydraulic Research, Vol. 1. Thomas Telford: London; pp. 510-515

17. Huang, H. Q. and G.C.,Nanson, 1998 : The influence of bank strength on channel geometry. Earth Surface Processes and Landforms, 23, pp. 865-876

18. Julien, P. Y. and J., Wargadalam, 1995 : Alluvial channel geometry : theory and applications. Journal of Hydraulic Engineering, ASCE, 121, pp. 312-325

19. Millar, R.G. and M.C. Quick, 1993 : Effect of bank stability on geometry of gravel rivers. Journal of Hydraulic Engineering, ASCE, 119, pp. 1343-1363.

20. Simons, D.B. and M.L. Albertson, 1960 : Uniform water conveyance channels in alluvial materials. Journal of the Hydraulics Division, ASCE, 86, pp. 33-71

21. Pickup, G., 1976 : Adjustment of stream channel shape to hydrologic regime. Journal of Hydrology, 30, pp. 365-373

22. Kirkby, M. J. 1977 : Maximum sediment efficiency as a criterion for alluvial channels. In River Channel Changes, Gregory, K. J. (ed.). Wiley : Chichester; pp. 429-442

23. White, W. R., Bettess, R. and Paris, E., 1982 : Analytical approach to river regime. Journal of the Hydraulics Division, ASCE, 108, pp. 1179-1193

24. Chang, H.H., 1979a: Geometry of rivers in regime, Journal of the Hydraulics Division, ASCE 105, pp. 691-706

25. Chang, H.H., 1979b: Minimum stream power and river channel patterns. Journal of Hydrology, 41, pp. 303-327

26. Chang, H.H., 1988 : Fluvial Processes in River Engineering. Wiley: New York

27. Chang, H.H., 1980a: Stable alluvial canal design. Journal of the Hydraulics Division, ASCE, 106, pp. 873-891

28. Chang, H.H., 1980b: Geometry of gravel streams. Journal of the Hydraulics Division, ASCE, 106, pp. 1443-1456

29. Chang, H.H., 1986 : River channel changes: adjustments of equilibrium. Journal of Hydraulic Engineering, ASCE, 112, pp. 43-55

30. Griffiths, G.A., 1984 : Extremal hypotheses for river regime: an illusion of progress. Water Resources Research, 20, pp. 113-118.

31. Yang, C.T., 1987 : Energy dissipation rate approach in river mechanics. In Sediment Transport in Gravel-Bed Rivers, Thorne, C.R., Bathurst, J. C., Hey, R.D. (eds). Wiley: Chichester; pp. 735-758

Chapter 9

HYDRAULIC FRACTURING IN FORMATIONS WITH PERMEABLE NATURAL FRACTURES

Olga Kresse[1] and Xiaowei Weng[1]
[1] Schlumberger, Sugar Land, USA

ABSTRACT

The recently developed Unconventional Fracture Model (UFM*) simulates complex hydraulic fracture network propagation in a formation with pre-existing closed natural fractures, and explicitly models hydraulic injection into a fracture network with multiple propagating branches [1]. The model predicts whether a hydraulic fracture front crosses or is arrested by a natural fracture it encounters, which defines the complexity of the generated complex hydraulic fracture network.

While taking into account the leakoff of the fracturing fluid into the formation, the leakoff into the natural fractures should also be considered, especially in low-matrix permeability conditions. The transmissibility of natural fractures can become significant, and the fracturing fluid can penetrate into natural fractures. Different regions can coexist along the invaded natural fracture: hydraulically opened region filled with fracturing fluid, region of still closed natural fracture invaded by fracturing fluid due to natural fracture permeability, and the region of natural fracture filled with original reservoir fluid.

Explicit modelling of hydraulic fractures interacting with permeable natural fractures becomes extremely complicated with the necessity to account for conservation of fluid mass, pressure drop along natural fractures, leak-off into the formation from natural fracture walls, pressure sensitive natural fracture permeability, properties of natural fractures, fluid rheology, while tracking the interface of each region along invaded natural fracture. A main challenge is integrating this hydraulic fracture/natural fracture interaction modelling into the overall hydraulic fracture network propagating scheme without losing model effectiveness and CPU performance.

The updated UFM model with enhancement to account for leakoff into the natural fractures will be presented.

INTRODUCTION

It is believed that complexity of the resulting fracture network during hydraulic fracturing treatments in formations with pre-existing natural fractures is caused mostly by the interaction between hydraulic and natural fractures. Natural fractures can be important for hydrocarbon production in the majority of low-permeability reservoirs, particularly where the permeability of the rock matrix is negligible. Understanding and proper modelling of the mechanism of hydraulic-natural fractures interaction is a key to explain fracture complexity and the microseismic events observed during HF treatments, and therefore to properly predict production.

When hydraulic fracture (HF) intercepts natural fracture (NF) it can cross the NF, open (dilate) the NF, or be arrested at NF. If hydraulic fracture crosses natural fracture, it remains planar, with a possibility to open the intersected NF if the fluid pressure at the intersection exceeds the effective stress acting on the NF. If the HF does not cross the NF, it can dilate and eventually propagate into the NF, which leads to more complex fracture network.

The interaction between HF and NF depends on in-situ rock stresses, mechanical properties of the rock, properties of natural fractures, and hydraulic fracture treatment parameters including fracturing fluid properties and injection rate. During the last decades, extensive theoretical, numerical, and experimental work has been done to investigate, explain, and use the rules controlling HF/NF interaction [3-17]. A new crossing model [2] recently implemented in UFM is able to predict the crossing behaviour of HF at the NF accounting for the effects of fluid properties and NF permeability [18].

One of the important effects of natural fractures is enhanced leakoff, which can lead to a premature screenout during proppant injection. In a formation with low-matrix-permeability, the transmissibility of natural fissures can be significantly higher than that of the reservoir matrix. The fracturing fluid can readily penetrate into natural fissures during the fracturing process and maintain a pressure nearly equal to the pressure in the primary fracture [19].

The concept that natural fractures (fissures) could alter leakoff has been a subject of numerous studies [20-24] with considering the fissure opening conditions or pressure-sensitive leakoff conditions. It was often reported that permeability of natural fractures is pressure dependent [23, 25, 26].

The ways that elevated pressure could affect natural fractures have been described in [27]. Fissures with rough surfaces and minimal mineralization are most likely highly sensitive to the net stress pushing on them. Under virgin reservoir conditions (when the pressure p within the fissure equals the initial reservoir pressure p_{ini}), the effective stress is fairly high and the open channels formed by mismatched fracture faces are most likely deformed and nearly closed. As the pressure in the fissure increases because of leakoff of the high-pressure fracturing fluid ($p > p_{ini}$), the net closure stress is reduced and the fissure porosity opens. In this regime, the leakoff coefficient is highly pressure dependent. As the pressure exceeds the closure stress on the fissure ($p > p_{fo}$), the entire fissure opens, yielding an accelerated leakoff condition. The estimation of the critical pressure in PKN-type HF (exceeding closing pressure $p > p_{fo}$) to open a vertical fissure has been given through the function of the principal horizontal stresses and Poisson ratio [20].

A more detailed description of the effects from natural fissures in reservoirs where natural fissures are the primary source of permeability is provided in [23]. The enhanced rate of fluid loss throughout the treatment is predicted, with leakoff accelerating as the fracturing pressure increases. The increase in fluid pressure in the fissures reduces the effective normal stress acting to close the fissures and hence increases their permeability. For hydraulic fracturing purposes, the effect of the magnified permeability is reflected as an increase in fluid-leakoff coefficient. The fluid-leakoff in the presence of natural fissures could be as high as 2 to 3 times that for normally occurring pressure – dependent leakoff behaviour, even under the net pressure conditions.

For slightly elevated pressures NF porosity begins to open as the pore pressure increases because the elevated pressure relieves some of the net stress on the asperity contacts. Several models of this process have been developed. For example, [26] predicts the change in NF permeability resulting from changes in stress and pressure. This model have been validated and used in numerous studies [23].

Among existing HF models accounting for the permeability of intercepted natural fractures mention [10] which couples fluid flow, elastic deformation, and frictional sliding to obtain a solution which depends on the competition between fractures for the permeability enhancements. The effect of initially closed but conductive fracture is specifically addressed. The possible scenarios for evolution of fracture opening and fluid transport in closed NFs implemented in [10] are shown in Figure 1. The initial aperture w_0 along a closed pre-existing NF corresponds to its residual conductivity. It is equal to the effective aperture for the parallel plate model. The initial conductivity of a closed natural fracture arises from the fact that its surfaces are rough and mismatched at fine scale, i.e. the aperture w_0 is related to the fracture porosity. With increasing the fluid pressure, the hydraulic aperture will slightly change due to micro structural change in the natural fracture, although the fracture still remain closed and carries some contact stresses. In the end, fracture will be opened mechanically as the fluid pressure exceeds the normal stress acting on the fracture. In this case, the effective hydraulic conductivity is equal to the sum of both hydraulic aperture and mechanical opening since the fracture opening augments the initial hydraulic aperture, as shown in Figure 1. Zhang's model also considers the possibility of frictional sliding through the Coulomb frictional law, and accounts for three types of contact behaviour at fracture surface: fracture is opened, fracture is closed but surface is in sticking mode, and fracture is closed but in sliding mode.

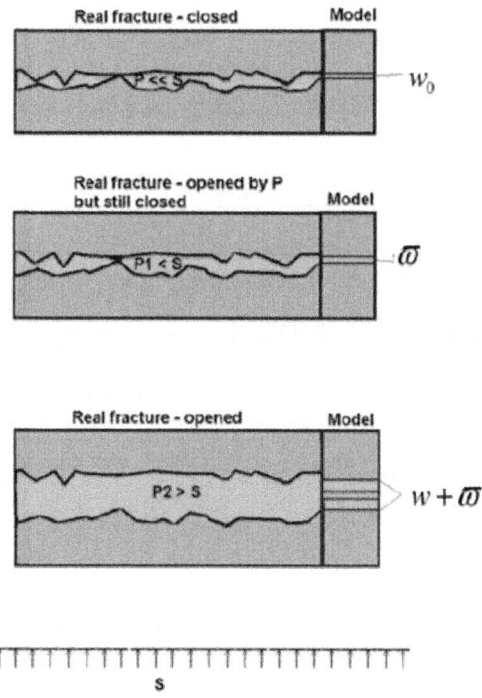

Figure 1. Evolution of natural fracture opening [10].

The HF models [28-32] do not account for permeability of natural fractures explicitly. The 2D model in [33] uses approach from [13] to simulate interaction between induced propagating fracture and natural fracture. A modified leak-off model for an intersecting fracture based on poro-elasticity was introduced to account for the increased leakoff at the intersections. A poro-elastic solution for the stresses in the HF/NF interaction zone has been used as a basis for hydraulic/natural fracture interaction criteria. A fully coupled finite element based approach was used to simulate HF propagation in a poroelastic formation with existing natural fractures.

The approach given in [10] is based on boundary element method and rigorously models HF interaction with permeable NF. It is computationally expensive, and is applicable for analysis of limited (small) number of HF/NF interactions. For a more general complex fracture network model like UFM which deals with a large number (order of thousands) of natural fractures, the CPU time is important and model should be computationally efficient while still being physically correct.

The important aspect of HF/NF interaction is shear slippage of NF faces. The possibilities of shear slippage in natural fractures due to change of stress field (in isolated natural fractures) or during HF/NF interactions, and the influence of shear slippage on fracture aperture change and dilation have been a subject of experimental and numerical studies [10, 15, 17, 30, 34-37]. The conditions for shear slippage and the corresponding shear displacement (apertures) have been investigated [36], and the estimation of permeability of NF with changing effective normal stress is done by [11,38]. The shear slippage effect during HF/NF interaction is also included in current approach.

This paper describes how leakoff into the natural fractures during HF/NF interaction (crossing or arresting before NF opens) is integrated into the complex hydraulic fracture model UFM.

UFM MODEL SPECIFICS

A complex fracture network model, referred to as Unconventional Fracture Model (UFM), had recently been developed [1, 39, 40]. The model simulates the fracture propagation, rock deformation, and fluid flow in the complex fracture network created during a treatment. The model solves the fully coupled problem of fluid flow in the fracture network and the elastic deformation of the fractures, which has similar assumptions and governing equations as conventional pseudo-3D fracture models. Transport equations are solved for each component of the fluids and proppants pumped. A key difference between UFM and the conventional planar fracture model is being able to simulate the interaction of hydraulic fractures with pre-existing natural fractures, i.e., determine whether a hydraulic fracture propagates through or is arrested by a natural fracture when they intersect and subsequently propagates along the natural fracture.

To properly simulate the propagation of multiple or complex fractures, the fracture model takes into account the interaction among adjacent hydraulic fracture branches, often referred to as "stress shadow" effect. It is well known that when a single planar hydraulic fracture is opened under a finite fluid net pressure, it exerts a stress field on the surrounding rock that is proportional to the net pressure. The details of stress shadow effect implemented in UFM are given in [40].

The branching of hydraulic fracture when intersecting natural fracture gives rise to the development of a complex fracture network. A crossing model that is extended from the Renshaw-Pollard [12] interface crossing criterion, applicable to any intersection angle, has been developed, validated against the experimental data [16, 17], and was integrated at first in the UFM. The crossing model, showing good comparison with existing experimental data, did not account for the effect of fluid viscosity and flow rate on the crossing pattern. More recently a new advanced OpenT crossing model, taking into the account the impact of fluid and NF properties, have been developed [2] and integrated in UFM [18].

The modelling approach used in UFM to predict the leakoff into the NFs is presented below.

MODELING LEAKOFF INTO PERMEABLE NF IN UFM

The main assumptions for the current modelling of leakoff from HF into the intercepted NF in UFM are given below.

- The rock formation contains vertical discrete deformable fractures (NFs), which are initially closed but conductive because of their pre-existing apertures (due to surface roughness, etc).
- The propagation direction of the hydraulic fractures is not affected (unless intercepted) by closed and not invaded natural fractures. The intercepted NF could affect HFN propagation even from closed parts (when shear slippage takes place)
- The natural fractures are assumed to contain pore space and are permeable.
- The original fluid inside the natural fractures and in the reservoir (oil, gas, water) is compressible and Newtonian.
- Fracturing fluid can be incompressible or compressible and its rheology can be Newtonian or power law.
- The rock material is assumed to be permeable and elastic.
- When intercepted by the main hydraulic fracture, a natural fracture may remain closed, while still being able to accept fracturing fluid, or may be mechanically opened by fracturing fluid pressure depending on the magnitude of the fracturing fluid pressure, confining stresses applied on natural fracture, and frictional properties of natural fracture.
- The flow inside NFs is assumed to be 1D.

- The natural fractures can be opened by fluid pressure that exceeds the normal stress acting on them and/or experience Coulomb type frictional slip.
- The original natural fracture has width w_0 and are filled with reservoir fluid with pressure equal to pore pressure $p_0 = p_{res}$.
- Fluid flow invaded into the natural fracture develops along NFs. Invaded fracturing fluid into the NF may reach the end of NF, break the rock, and start to propagate into the rock accordingly to previously implemented propagation rules (only if the NF is opened). Fracture re-initiation from other points along the NF other than its ends (offsets) is not modeled at this time.

Hydraulic fractures propagating in the rock are modeled in accordance with existing approach in UFM model. The Schematic of the complex HF interaction with permeable NF is shown in Figure 2 and Figure 4.

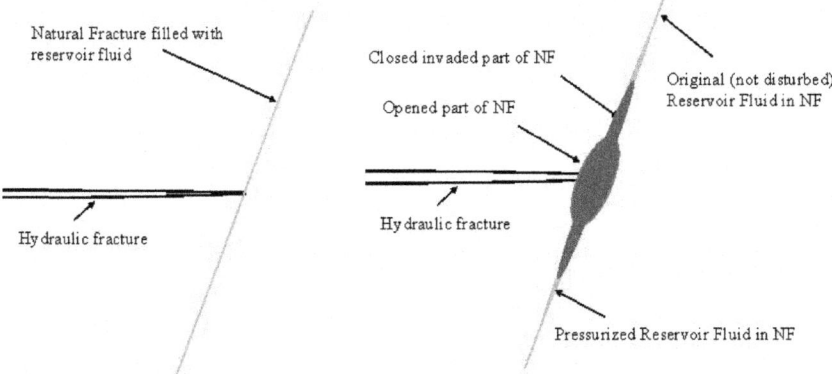

Figure 2. Hydraulic fracture intercepting natural fracture and possible situation to model.

Fracturing fluid invasion into the two wings of the NF needs to be considered separately. Four possible regions can co-exist in each wing of the NF encountered by HF (Figure 4):

- Opened part filled with invaded fracturing fluid (fluid pressure exceeds the normal effective stress on NF), with length of opened part $L_{opened} > 0$
- Invaded closed part of NF (filtration zone) filled with fracturing fluid (fluid pressure above pore pressure but below the closure stress) with length $L_{filtration} > 0$
- Closed pressurized part filled with pressurized original reservoir fluid (fluid pressure above the pore pressure) with length $L_{pressurized} > 0$

- Closed undisturbed part of NF filled with reservoir fluid under original pore pressure conditions.

When a natural fracture is intercepted by the hydraulic fracture, the fluid pressure in the hydraulic fracture transmits into the natural fracture. If the fluid pressure is less than the normal effective stress on the natural fracture, the natural fracture remains closed. Even closed natural fractures may have hydraulic conductivities much larger than the surrounding rock matrix, and in this case fracturing fluid will invade the natural fractures more than leakoff into the surrounding matrix. If the portion of injected fluid is lost into closed natural fractures from the main HF, the HF growth could be affected.

For a closed fracture, the equivalent fluid conductivity is expected to change with the fluid pressure since contact deformation is a function of effective normal stress. This pressure-induced dilatancy and the associated increase in conductivity are important in increasing leakoff. Also, any reduction in effective contact stress may result in fracture sliding, which can lead to local stress variations and slip induced fracture dilation, which can in turn change the overall conductivity of fracture networks.

The governing processes in first three regions listed above should be modeled, and the modeling approaches in different regions (also referred to as zones in the following context) are different due to different flow behaviors and rock/fluid properties.

BASIC GOVERNING EQUATIONS

Continuity of Fluid Volume (MASS)

The equation for the continuity of incompressible fluid volume has the form

$$\frac{\partial q_{NF}}{\partial s} + \frac{\partial A}{\partial t} + q_L = 0, \qquad q_L = \frac{2hC_{tot}^{rock}}{\sqrt{t - \tau(s)}}, \qquad A = \varpi h \tag{1}$$

where

$q_{NF}(t)$ - volumetric flow rate through a cross section of area A of natural fracture [m³/s]

A - cross sectional area of the natural fracture

q_L - the volume rate of leakoff per unit length

ϖ - average hydraulic fracture width (different from w, fracture opening by fluid pressure exceeding normal stress)

h - fracture height

C_{tot} - total leakoff coefficient from the wall of natural fracture

More generally in the case of compressible fluid the equation (1) should account for fluid density ρ_f and mass flux q_m. Considering the rate of change of fluid mass per unit length in a fracture m, continuity of fluid mass in the fracture is governed by equation

$$\frac{\partial q_m}{\partial s} + \dot{m} + \rho_f q_L = 0$$

(2)

or along the fracture of constant length

$$\frac{\partial q_m}{\partial s} + \frac{\partial(\rho_f \varpi h)}{\partial t} + \rho_f q_L = 0, \qquad q_L = \frac{2hC_{tot}^{rock}}{\sqrt{t - \tau(s)}}$$

(3)

Here s is coordinate along NF, and total leakoff coefficient from the walls of the natural fracture C_{tot}^{rock} is equal to combined leakoff coefficient [41]

$$C_{tot}^{rock} = C_{vc}^{rock} = \frac{2C_v^{rock}C_c^{rock}}{C_v^{rock} + \sqrt{C_v^{rock2} + 4C_c^{rock2}}}$$

(4)

Where leakoff coefficient for the filtration zone in the rock and leakoff coefficient for the reservoir zone as shown in equations (5a - 5b)

$$C_v^{rock} = \sqrt{\frac{k_r \varphi_r \Delta p}{2\mu_f}} = \sqrt{\frac{k_r \varphi_r \Delta p}{2\mu_f}}, \Delta p = p_f - p_s \text{a}$$

$$C_c^{rock} = \sqrt{\frac{k_r \varphi_r c_T}{\pi \mu_r}} \Delta p, \qquad \Delta p = p_f - p_r \text{b}$$

$$\overline{C}_v = \frac{2(C_v)^2 \sqrt{t}}{V_L} \text{c}$$

(5)

with

ϕ_r - reservoir porosity

c_T - total compressibility of reservoir

k_r - permeability of rock matrix

μ_r - reservoir fluid viscosity in the porous media

μ_f - filtrate fluid viscosity

ρ_f - filtrate fluid density

p_r - reservoir pressure

In the case of multiple fluids, the invaded zone can be described by replacing C_v with equivalent term (see (5c)) [42] Where $\overline{C}_v$ is calculated using the average viscosity and relative permeability of all the filtrate fluids leaked off up to the current time, and V_L is the fluid volume per unit area that previously leaked off into the reservoir.

Pressure Drop along Closed NF

The pressure drop along closed NF can be expressed from Darcy's law

$$q_{NF} = -\frac{k_{NF} A}{\mu_f} \frac{\Delta p}{L(t)}, A = \varpi h$$

(6)

Or for mass flux

$$q_m = -\rho_f \frac{k_{NF}}{\mu_f} A \frac{\partial p}{\partial s}$$

(7)

Then the pressure can be calculated from

$$\frac{\partial p}{\partial s} = -\frac{\mu_f}{\rho_f k_{NF} A} q_m = -\frac{\mu_f}{\rho_f k_{NF} \varpi h} q_m, \quad \text{at the inlet} : p = p_{in}(t)$$

(8)

Here

k_{NF}- permeability of natural fracture

μ_f- filtrate fluid viscosity

ρ_f- filtrate fluid density

A - cross sectional area of closed NF

p_{in}- fluid pressure at the inlet

Change of NF Permeability Due To Stress and Pressure Changes

The leakoff into the natural fracture, or permeability of the natural fracture, is highly pressure dependent when pressure of invading fluid exceeds reservoir pressure but still is below the closure pressure. In general, permeability of natural fracture is a function of normal stress on NF, shear stress (or shear displacement due to shear slippage), and fluid pressure, and can be represented as a combination of permeability due to normal stress and permeability due to shear slippage [26]:

$$k_{NF} = f(k_{NF}^n, k_{NF}^s)$$
$$k_{NF}^n = f_1(k_o, \sigma_n, p), \quad k_{NF}^s = f_2(u_s, \phi_{dil})$$
$$k_{NF}^n = k_o \left\{ C \ln \left[\frac{\sigma^*}{\sigma_n - p} \right] \right\}^3$$

(9)

Where constants C and σ^* (reference stress state) are determined from field data, k_o is the initial NF permeability (reservoir permeability under in-situ conditions), σ_n is the normal stress on the NF, p is the pressure in the NF, and u_s is shear-induced displacement (slippage).

The Width of Closed Invaded NF

The width ϖ of closed NF invaded by the treatment fluid (hydraulic aperture) is related to the pressure-dependent permeability as [10]

$$\varpi = \sqrt{12 k_{NF}}$$ (10)

The hydraulic width ϖ can be evaluated from Barton-Bandis model following approach [11,36,38], i.e. directly from Eq. (11) for given effective normal stress σ_{eff}, reference effective stress σ_n^{ref}, initial hydraulic fracture aperture ϖ_o (related to the roughness of fracture surface), shear displacement u_s and dilation angle ϕ_{dil}

$$\varpi = \frac{\varpi_0}{1+9\frac{\sigma_{eff}}{\sigma_n^{ref}}} + \varpi_s + \varpi_{res}, \quad \varpi_s = \left| u_s \right| \tan(\phi_{dil}^{eff}), \quad \phi_{dil}^{eff} = \frac{\phi_{dil}}{1+9\frac{\sigma_{eff}}{\sigma_n^{ref}}}$$

$$\sigma_{eff} = \sigma_n - p_f(s)$$ (11)

Shear Failure

The second term in Eq. (11) represents the shear-induced dilation which contributes to the NF permeability (Eq.9). Shear induced dilation is related to the frictional slip which occurs when the shear stress reaches the frictional shear strength of the natural fractures $\tau_s = \lambda(\sigma_n - p)$. In the present study the NF propagation due to the shear induced slip is not considered, but the contribution of shear slippage zone to the NF enhanced permeability is evaluated based on the enhanced 2D DDM approach [40,43,44].

$$\sigma_n^i - p^i = \sum_j A^{ij} C_{nn}^{ij} D_n^j + \sum_j A^{ij} C_{ns}^{ij} D_s^j$$

$$\tau^i = \sum_j A^{ij} C_{sn}^{ij} D_n^j + \sum_j A^{ij} C_{ss}^{ij} D_s^j, \quad A^{ij} = 1 - \frac{d_{ij}^{2.3}}{\left(d_{ij}^2 + h^2\right)^{2.3/2}}$$ (12)

The fracture surface slip (shear displacement) u_s can be found as shear displacement discontinuity D_s calculated for the closed sliding elements following Coulomb frictional law $\tau \geq \tau_s = \lambda(\sigma - p)$ from elasticity equations in the stress shadow calculation approach with accounting for the mechanical opening in HFN from Eq. (12).

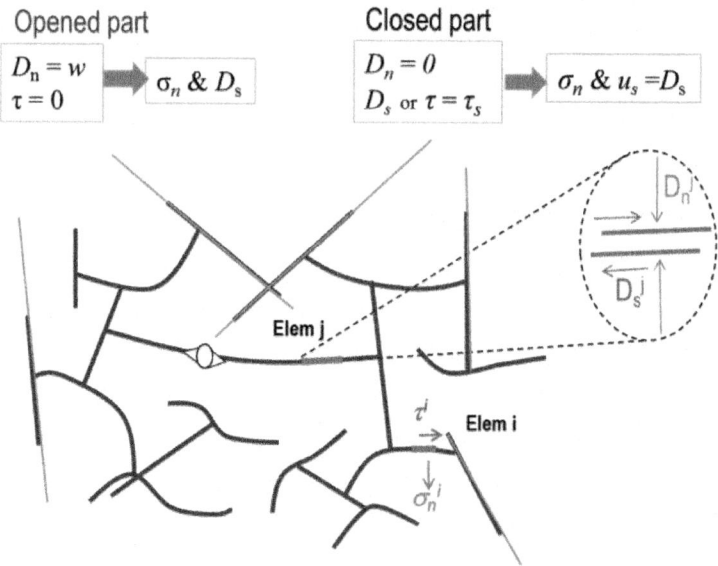

Figure 3. Stress Shadow Effect from opened (blue) HFN and closed parts (grey) of intercepted NFs

For any element j in the opened part of HFN (including opened part of intercepted NFs) the input normal displacement discontinuity D_n^j in Equation (12) is given by known fracture aperture (width) $D_n^j = w^j \neq 0$, and the shear stress is zero $\tau^j = 0$. Along the closed part of intercepted NFs the mechanical the opening is zero, $D_n^j = 0$, and the pressure and normal stress should be tracked to detect (find) elements sliding in shear. If $\tau^i = \tau_s^i$ then element i is sliding and dilating in shear, and the shear stress $\tau^i = \tau_s^i$ for this element is used to find fracture surface slip $u_s^i = D_s^i$ from the second equation (12). So, equation (12) could be solved for shear displacements D_s^j along opened and sliding in shear parts of total HFN and intercepted NF as schematically shown in Fig.3. In Eq.(12) C^{ij} are 2D, plane strain elastic influence coefficients [43] defining interactions between the elements i and j, and A^{ij} are 3D correction factors [44] accounting for the 3D effect due to fracture height h depending on the distance between elements d_{ij}.

Fluid Density as Function of Pressure and Temperature

Density of gas as a function of pressure and temperature has form

$$\rho_{gas} = \frac{p m_m}{Z R T}$$
(13)

Where p is pressure, T is temperature, Z is compressibility, R is gas constant, and m_m is molar mass. The changes in pressure and temperature with time produce changes in gas density. Density of a compressible fluid as function of pressure

$$\frac{d\rho}{dt} = \frac{\rho}{B}\frac{dp}{dt} = \rho c_f \frac{dp}{dt}$$
(14)

where B is bulk modulus (fluid elasticity) in Pa and $c_f = 1/B$ is fluid compressibility in Pa^{-1}. More generally, change in fluid density due to changes in pressure and temperature for the low compressibility fluids through the known values of density ρ_0 and temperature T_0 at pressure p_0 has form (β here is volumetric expansion coefficient)

$$\rho = \frac{\rho_0}{1-\beta(T - T_0)} \times \frac{1}{1 - \frac{p - p_0}{B}}$$
(15)

Fluid Flow in Opened NF

The fluid flow in the opened NF ($p f > \sigma n$) will be handled as fluid flow in HFN and have been described before [1] depending on the flow regime:

Laminar fluid flow: Poiseuille Law [45]

$$\frac{\partial p}{\partial s} = -\alpha_0 \frac{1}{\bar{w}^{2n'+1}} \frac{q}{H_{fl}} \left| \frac{q}{H_{fl}} \right|^{n'-1}$$
(16)

$$\alpha_0 = \frac{2K'}{\varphi(n')^{n'}} \left(\frac{4n'+2}{n'} \right)^{n'}; \quad \varphi(n') = \frac{1}{H_{fl}} \int_{H_{fl}} \left(\frac{w(z)}{\bar{w}} \right)^{\frac{2n'+1}{n'}} dz$$

Where $\bar{w}$ is average fracture opening, and n' and K' are fluid power law exponent and consistency index.

Turbulent fluid flow ($N_{Re} > 4000$):

$$\frac{\partial p}{\partial s} = -\frac{f \rho_f}{\bar{w}^3} \frac{q}{H_{fl}} \left| \frac{q}{H_{fl}} \right|, q = H_{fl} \left(-\frac{\bar{w}^3}{f \rho_f} \frac{dp}{ds} \right)^{1/2}$$
(17)

With Reynolds number (N_{Re}) for the power law fluid between parallel plates and *Fanning* friction factor (f) defined as

$$N_{Re} = \frac{3^{1-n'} 2^{2-n'} \rho V^{2-n'} \bar{w}^{n'}}{K' \left(\frac{2n'+1}{3n'} \right)^{n'}}, f \approx \frac{1}{16 \left[\log_{10} \left(\frac{\varepsilon}{7.4\bar{w}} \right) \right]^2};$$
(18)

Where V is fluid velocity and ε is surface roughness height.

Darcy fluid flow through proppant pack of height h

$$\frac{\partial p}{\partial s} = -\frac{q\mu_{fl}}{kh\overline{w}}$$

(19)

will take place if the height of the fluid in fracture element become smaller than minimum fluid height. The minimum fluid height is calculated from the condition that pressure drop is equal to pressure drop due to the Darcy flow. The minimum height for turbulent and laminar flow is defined as

Laminar flow: $\quad h_{fl}^{min} = \left(\frac{k\alpha_0(n')H_{fl}}{\overline{w}^{2n'}\mu_{fl}}\right)^{1/n'} \frac{1}{q^{\frac{1}{n'}-1}}$

Turbulent flow: $\quad h_{fl}^{min} = \left(\frac{kH_{fl}f\rho q}{\overline{w}^2\mu_{fl}}\right)^{1/2}$

(20)

Where μ_{fl} is fluid dynamic viscosity, and k is proppant pack permeability. The boundary conditions at the inlet and tip of opened fracture

$$p = p_f(t), \quad p_{tip} = \sigma_n^{tip}(t)$$

(21)

where p_f is known fluid pressure at the intersection with natural fracture.

COMBINED FLUID FLOW INTO THE OPENED AND CLOSED PARTS OF INVADED NF

As I mentioned before, natural fracture can be closed, closed but invaded with fracturing fluid, closed and filled with pressurized reservoir fluid, or opened (Figure 4). The partially opened NF can contain opened, invaded, pressurized and closed parts which are dynamically changing with time. When fronts (positions of the boundaries between co-existing parts in invaded NF) change or propagate, the velocity of each propagating front can be considered as velocity of the corresponding fluid front in the relevant part of the invaded NF.

To properly model invasion of fracturing fluid into the NF, the propagation of each front should be modelled, and in different parts of the invaded NF different governing equations should be satisfied.

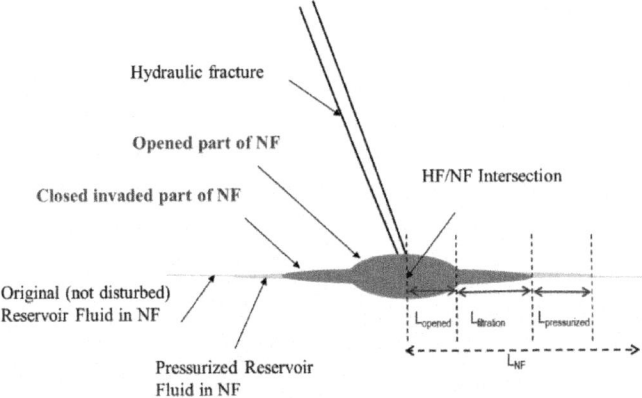

Figure 4. Details on different possible zones in intercepted permeable NF.

Opened Part OF NF

If the NF is opened at intersection element i with HF, then fluid pressure at intersection exceed the local normal stress on NF:

$$p_f(i) > \sigma_n^{NF}(i) \tag{22}$$

The tip of opened part of NF or the intersection element with HF (if NF is closed) becomes the inlet (injection point) into the closed invaded part of NF (filtration zone) $p_{in}^{filtr} = p_{tip}^{open} = \sigma_n^{NF}$. Mention that if NF is completely opened, then it is a part of total hydraulic fracture network (HFN) and handled based on the approach described previously in [1], see also equations (16)-(21).

If NF is closed at intersection element i with HF, then fluid pressure is below the local normal stress, but can still be higher than reservoir (pore) pressure

$$p_o < p_f(i) \le \sigma_n^{NF}(i) \tag{23}$$

Filtration Zone (Closed Part of NF Invaded By Fracturing Fluid)

The fluid pressure, width and flow rate along the filtration zone can be calculated from given pressure $p_{in}^{filtr} = p_f(i)$ which satisfies condition (23). The flow rate along filtration zone can be iteratively solved from the system of equations (24) with flow rate from the inlet to filtration zone q_{in}^{filtr} which is a part of total solution

$$
\begin{cases}
\dfrac{\partial p}{\partial s} = -\dfrac{\mu_f}{k_{NF}\varpi_{filtr}h}q, & \text{at the inlet}: \ p = p_{in}^{filtr}(t) \\[3mm]
\dfrac{\partial q}{\partial s} + \dfrac{\partial(\varpi_{filtr}h)}{\partial t} + q_L = 0, \ q_L = \dfrac{2hC_{tot}^{rock}}{\sqrt{t-\tau_o(s)}} \\[3mm]
k_{NF} = \dfrac{\varpi_{filtr}^2(p)}{12} \\[3mm]
\varpi_{filtr}(p) = \dfrac{\varpi_0}{1+9\dfrac{\sigma_{eff}}{\sigma_n^{ref}}} + \varpi_s + \varpi_{res}, \ \ \varpi_s = |u_s|\tan(\phi_{dil}^{eff}), \ \ \sigma_{eff} = \sigma_n - p(s),
\end{cases}
\tag{24}
$$

The length of filtration zone can be calculated by the tracking the volume of fracturing fluid leaked into the NF by marching from the inlet along the NF. At the end of filtration zone mass balance should be satisfied and fluid pressure will be higher or equal to the reservoir pressure

$$
q_{in}^{filtr}\,dt = dVol_{frac}^{filtr}(dt) + dVol_{leak}^{filtr}(dt)
$$

$$
p_r < p^{endfiltrzone} < \sigma_n^{NF}
\tag{25}
$$

The flowrate (filtration/pressurized front velocity) at the last element of filtration zone is used for calculations in the pressurized zone. The position of the front between the filtration zone and pressurized zone should be tracked, giving velocity of filtration front and pressure p_{in}^{pres} as input for solution in pressurized zone.

If NF is partially opened, then in the solution scheme for the filtration zone the intersection (inlet) element is replaced by the tip element i of the opened part of NF with $p(i) = \sigma_n^{NF}(i)$.

Closed pressurized part of NF (filled with reservoir fluid).

Mention that if the leakoff coefficient for filtration zone $C_v^{rock} = 0$, then $p_r = p_f^{end\ filtr\ zone} < \sigma_n^{NF}$, and there will not be pressurized zone in NF ($p_r = p_f^{end\ filtr\ zone} < \sigma_n^{NF}$, $L_{pressurized} = 0$).

For the general case of compressible reservoir fluid and non-zero total leakoff from the walls of pressurizes NF into the rock, the leakoff to the rock from the NF part filled with pressurized reservoir fluid is defined by the compressibility controlled leakoff coefficient

$$
C_c^{rock} = \sqrt{\dfrac{k_r\varphi_r c_T}{\pi\mu_r}}\,\Delta p, \quad \Delta p = p_{NF} - p_r
\tag{26}
$$

The governing equations to calculate fluid pressure, width, and flow rate along the pressurized zone from known influx $q_{filtr/pres}$ and pressure $p_{filtr/pres}$ are similar to Equations (24) for filtration zone with replacing fracturing fluid with reservoir fluid and using ϖ_{pres} as hydraulic width of pressurized zone of invaded NF, and $c\,T$ as reservoir fluid compressibility.

Notice that the pressure at the front between opened and filtration zones along invaded NF (or at HF/closed NF intersection point) and the time step are the inputs for the new pressure, width, and flow rate calculation in closed invaded part of intercepted NF. Initial influx q_{in} can be prescribed based on the pressure in NF from the previous time step, and then the solution scheme will be applied from the intersection towards the end of NF with tracking the incremental mass balance (fluid injected to NF at current time step) to define the end of filtration front, and tracking pressure in the rest of NF (not invaded part, filled with reservoir fluid) to track the end of pressurized zone (fluid pressure equal to reservoir pressure). The flow rate and pressure are tracked and corresponding front positions are to be updated iteratively until in the pressurized zone of NF the following condition is satisfied (which indicates the position of the end of the pressurized zone)

$$q(L_{open} + L_{filtr} + L_{pres}) = 0$$
$$p(L_{open} + L_{filtr} + L_{pres}) = p_r \tag{27}$$

At the end of pressurized zone pressure is equal to the reservoir pressure and flow rate is zero. If the end of NF is reached and pressure is above the reservoir pressure, then Equation (27) is replaced with condition $q(L_{NF})=0$. During pumping the pressurized zone extends until the end of NF, and then gradually shrinks, while the lengths of invaded zone and opened zones increase. Eventually whole NF will be filled with fracturing fluid, so the NF will contain only filtration and/or opened zones. When NF is completely opened, it becomes a part of total HFN. The elements in closed part of invaded NF can slip under some conditions (for example, due to the stress field change from stress shadow), influencing the calculations of total NF permeability k_{NF}.

Each element in the closed part of intercepted NF is checked for shear slip possibility. The fracture surface slip (the shear displacement u_s) can be found as shear displacement discontinuity calculated for the closed sliding elements satisfying Coulomb frictional law $\tau \geq \tau_s = \lambda(\sigma - p_f)$, from elasticity equations (12) accounting for the mechanical opening in HFN.

If some elements in closed not disturbed part of NF are sliding then the corresponding shear stress is involved in the stress shadow calculations and thus influences simulations results.

NUMERICAL APPROACH DESCRIPTION

The treatment of permeable natural fractures is a part of UFM model. It is possible to model leakoff from HF into the NF form UFM in different ways, depending on the importance of required accuracy, numerical stability and CPU time.

The most computationally expensive but at the same time most accurate is a fully coupled numerical approach. The fully coupled approach means to discretise different parts of the invaded NFs and numerically solve the pressure as a part of total system of equations for the fracture network using iterative solution scheme. Another, more CPU efficient approach, is a decoupled numerical approach, where pressure and width along the invaded but closed part of NF are calculated separately based on the results from previous time step calculations along the HFN and corresponding pressure at HF/NF intersections (inlets).

Two approaches have been considered to model interaction of hydraulic fracture with permeable natural fractures.

Decoupled Numerical Approach

- When NF is intercepted by HF, create elements along whole NF to be used at next time step
- Evaluate initial guess of flow rate into the NF based on the pressure at the intersection and old pressure profile along closed NF
- Check for a possibility of frictional sliding along the closed parts of NFs to evaluate pressure-dependent permeability and conductivity along NF
- Iteratively calculate pressure, hydraulic width, flow rate and length of each zone along NF with using corresponding equations (for filtration and pressurized zones) by marching from intersection till the end of NF until condition (27) is satisfied indicating final results and final positions of zones' fronts
- Track the end of filtration zone for each pressure and flow rate iterations by checking the volume of fluid injected into NF at given time step
- Save invaded volume for volume balance, influx for mass balance, pressure at intersection, and time step to be used at the next time step
- Track the pressure at intersections and/or tips of opened HF part in NF to capture opened zones.
- If intersection is opened, use the pressure at the tip of opened HFN part along invaded NF for calculations along corresponding closed NF part
- Apply rules to treat special situations (intersecting NFs, etc)
- Save elements information (volume, pressure, flow rates) for the next time step

Fully Coupled Numerical Approach

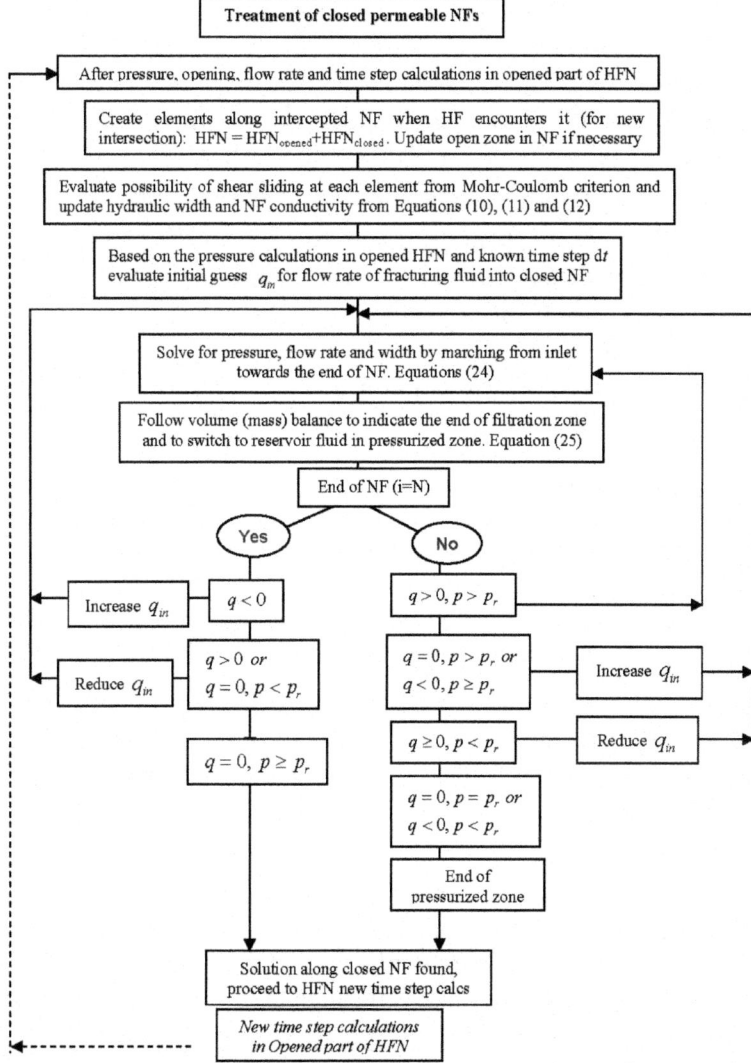

Figure 5. The proposed algorithm to account for leakoff from HF into the permeable NFs ($i=N$ indicates the end of NF).

- Discretize the NF when HF intercepts it
- Make the NF elements a part of total network (HFN) and include pressure calculations along different zones of NF into the whole iterative scheme to calculate pressure, time step, and front positions.

- Calculate the flow rate and volume of fluid injected to NFs at each pressure iteration, and update/calculate pressure along HFN using existing scheme for opened elements and new equations (described above) for elements in closed invaded and pressurized NF parts
- From the calculated pressure in NF elements iteratively update the positions of the propagating fronts for opened, filtration, and pressurized zones in NF during pressure/time step iterations
- Track the pressure at intersections to capture when NF start to open
- Include the elements in the closed parts of invaded NFs into the stress shadow calculation scheme

The fully coupled NF modeling approach is heavier and more CPU expensive than decoupled approach. The Decoupled Numerical Approach has been selected as basic approach and it is described schematically on Figure 5 as a part of total solution

CONCLUSIONS

The new approach developed to account for the complex processes due to NF permeability accompanying HF/NF interaction have been presented in detail. This approach accounts for the important physical processes taking place during HF and permeable NF interaction, and will be implemented in UFM.

The next step is to evaluate the influence of leakoff into the natural fractures during HFN simulations on the total HFN footprint and production forecast. The approach will be also validated against existing numerical, experimental and field data.

REFERENCES

1. O Kresse, C. E Cohen, X Weng, R Wu, H Gu, Numerical modeling of hydraulic fracturing in naturally fractured formations. 45th US Rock Mechanics/ Geomechanics Symposium, San Francisco, CA, 26 29June, 2011

2. D Chuprakov, O Melchaeva, R Prioul, Hydraulic Fracture Propagation Across a Weak Discontinuity Controlled by Fluid Injection. InTech; 2013

3. D. S Chuprakov, A. V Akulich, E Siebrits, M Thiercelin, Hydraulic Fracture Propagation in a Naturally Fractured reservoir. SPE 128715, Presented at the SPE Oil and Gas India Conference and Exhibition held in Mumbai, India, 20 22January, 2010

4. J Zhao, M Chen, Y Jin, G Zhang, Analysis of fracture propagation behavior and fracture geometry using tri-axial fracturing system in

naturally fractured reservoirs. Int. J. Rock Mech. & Min. Sci, 45 2008 1143 1152

5.   M Thiercelin, E Makkhyu, Stress field in the vicinity of a natural fault activated by the propagation of an induced hydraulic fracture. Proceedings of the 1st Canada-US Rock Mechanics Symposium 2007

6.   X Zhang, R. G Jeffrey, The role of friction and secondary flaws on deflection and re-initiation of hydraulic fractures at orthogonal pre-existing fractures. Geophysical Journal International 2006 166 3 1454 1465

7.   X Zhang, R. G Jeffrey, Reinitiation or termination of fluid-driven fractures at frictional bedding interfaces. J Geophys Res-Sol Ea. 2008Aug 28;113(B8).

8.   X Zhang, R. G Jeffrey, M Thiercelin, Effects of Frictional Geological Discontinuities on hydraulic fracture propagation. SPE 106111, Presented at the SPE Hydraulic Fracturing Technology Conference, College Station, Texas, January29 31 2007

9.   X Zhang, R Jeffrey, M Thiercelin, Deflection and propagation of fluid-driven fractures as frictional bedding interfaces: a numerical investigation. Journal of Structural Geology 2007

10.   X Zhang, R. G Jeffrey, M Thiercelin, Mechanics of fluid-driven fracture growth in naturally fractured reservoirs with simple network geometries. Journal of Geophysical Research 2009B12406.

11.   Beugelsdijk LJLde Pater CJ, Sato K. Experimental hydraulic fracture propagation in a multi-fractured medium. SPE 59419, presented at the SPE Asia Pacific Conference in Integrated Modeling for Asset Management, Yokohama, Japan, April 25 26 2000

12.   C. E Renshaw, D. D Pollard, An Experimentally Verified Criterion for Propagation across Unbounded Frictional Interfaces in Brittle, Linear Elastic-Materials. International Journal of Rock Mechanics and Mining Sciences & Geomechanics Abstracts. 1995Apr, 32 3 237 49

13.   N. R Warpinski, L. W Teufel, Influence of Geologic Discontinuities on Hydraulic Fracture Propagation (includes associated papers 17011 and 17074). SPE Journal of Petroleum Technology. 1987 39 2 209 220

14.   T. L Blanton, An Experimental Study of Interaction Between Hydraulically Induced and Pre-existing Fractures. SPE 10847, Presented at the SPE/ DOE Unconventional Gas Recovery Symposium, Pittsburgh, PA, May 16 18 1982

15.   T. L Blanton, Propagation of Hydraulically and Dynamically Induced

Fractures in Naturally Fractured Reservoirs. SPE Unconventional Gas Technology Symposium; 01/01/1986; Louisville, Kentucky 1986

16. H Gu, X Weng, Criterion For Fractures Crossing Frictional Interfaces At Non-orthogonal Angles. 44th US Rock Mechanics Symposium and 5th US-Canada Rock Mechanics Symposium; 01/01/2010; Salt Lake City, Utah: American Rock Mechanics Association; 2010

17. H Gu, X Weng, J. B Lund, M Mack, U Ganguly, R Suarez-rivera, Hydraulic fracture crossing natural fracture at non-orthogonal angles, a criterion, its validation and applications. Paper SPE 139984 presented at the SPE Hydraulic Fracturing Conference and Exhibition, Woodlands, Texas, 24 26January, 2011

18. O Kresse, X Weng, D Chuprakov, R Prioul, C. E Cohen, Effect of Flow Rate and Viscosity on Complex Fracture Development in UFM. InTech; 2013

19. M. J Economides, K. G Nolte, Reservoir Simulation. Third edition. 2000

20. K. G Nolte, M. B Smith, Interpretation of Fracturing Pressures SPE 8297, September 1981

21. J. L Castillo, Modified Fracture Pressure decline Analysis Including Pressure-Dependent leakoff, SPE 16417, 1987

22. K. G Nolte, Fracturing Pressure Analysis for Non-Ideal Behavior. SPE 20704, JPT, February 1991

23. N. R Warpinski, Hydraulic Fracturing in Tight, Fissured Media. SPE 20154, JPT, February 1991

24. R. D Barree, H Mukherjee, Determination of Pressure Dependent Leakoff and its effect on Fracture geometry, SPE 36424, 1996. Presented at the 71st Annual Tech Conference and Exhibition, Denver Co, 6 9October, 1996

25. H Mukherjee, S Larkin, W Kordziel, Extension of Fracture Pressure Decline Curve Analysis to Fissured Formations. SPE 21872, 1991. Presented at the Rocky Mountain Regional meeting and Low Permeability Reservoirs Symposium, Denver, Co, April 15 17 1991

26. J. B Walsh, Effect of Pore Pressure and Confining Pressure on Fracture Permeability, In. J. Rock Mech. Min. Sci. & Geomech. Abstr. 1981 18 429 435

27. N. R Warpinski, Fluid leakoff in natural fissures. In: Economides&Nolte: Reservoir Stimulation, 2000

28. B. R Meyer, L Bazan, A Discrete Fracture Network Model for Hydraulically Induced Fractures: Theory, Parametric and Case Studies,

SPE 140514. Presented at the SPE Hydraulic Fracturing Tech Conference and Exhibition in The Woodlands, Texas, 2426 January 2011

29.   S Rogers, D Elmo, R Dunphy, D Bearinger, Understanding Hydraulic fracture geometry and interactions in the Horn River Basin through DFN and Numerical modeling, SPE 137488, 2010. Presented at the Canadian Unconventional Resources & International Petroleum Conference, Calgary, Alberta, Canada, 19 21October, 2010

30.   N Nagel, B Damjanac, X Garcia, M Sanchez-nagel, Discrete Element Hydraulic Fracture Modeling- Evaluating Changes in natural Fracture Aperture and Transmissivity, SPE 148957, 2011. Presented at the Canadian Resources Conference, Calgary, Alberta, Canada, Novebmer 15 17 2011

31.   P Fu, S. M Johnson, C. R Carrigan, Simulating Complex Fracture Systems in Geothermal Reservoirs Using an Explicitly Coupled Hydro-Geomechanical model, ARMA 11 244Presented at 45th US Rock Mechanics/Geomechanics Symposium, Salt Lake City, UT, June 27-29, 2011

32.   W. S Dershowitz, M. G Cottrell, D. H Lim, T. W Doe, A Discrete Fracture Network Approach for Evaluation of Hydraulic Fracture Stimulation of Naturally Fractured Reservoirs, ARMA-475, 2010. Presented at 44th US Rock Mechanics Symposium, San Francisco, CA, June 26 29 2010

33.   M. M Rahman, A Aghigi, A. R Sheik, Numerical Modeling of Fully Coupled Hydraulic Fracture propagation in Naturally Fractured Poro-Elastic Reservoirs", SPE 121903, 2009. Presented at the 2009 SPE EUROPEC/EAGE Conference, Amsterdam, The Netherlands, 8 11June 2009

34.   S. R Brown, R. L Bruhn, Fluid permeability of deformable fracture network. Journal of Geophys. Research 1998B2): 2489-2500.

35.   M. L Cooke, C. A Underwood, Fracture termination and step-over at bedding interfaces due to frictional slip and interface opening. Journal Structural Geology,2001 23 223 238

36.   M. M Hossain, M. K Rahman, S. S Rahman, Volumetric Growth and Hydraulic Conductivity of naturally fractured reservoirs during hydraulic fracturing: A case study using Australian Conditions, SPE 63173, 2000. Presented at the 2000 SPE Technical Conference and Exhibition, Dallas, Texas, 1 4October 2000

37.   D. A Chuprakov, A. V Akulich, E Siebrits, M Thiercelin, Hydraulic-Fracture Propagation in a Naturally Fractured Reservoir.SPE 128715,

2011. Presented at the SPE Oil and Gas India Conference and Exhibition, Mumbai, India, 20 22January 2010

38. K Tezuka, T Tamagawa, K Watanabe, Numerical Simulation of Hydraulic Shearing in Fractures Reservoir. Proceeding World Geothermal Congress, Antalya, Turkey, 24 25April 2005

39. X Weng, O Kresse, C Cohen, R Wu, H Gu, Modeling of Hydraulic Fracture Network Propagation in a Naturally Fractured Formation. Paper SPE 140253 presented at the SPE Hydraulic Fracturing Conference and Exhibition, Woodlands, Texas, USA, 24 26January, 2011

40. O Kresse, X Weng, R Wu, H Gu, Numerical modeling of Hydraulic fractures interaction in complex Naturally fractured formations. ARMA-292, Presented at 46th US Rock Mechanics /Geomechanics Symposium, Chicago, Il, USA, 24 27June 2012

41. R. H Dean, S. H Advani, An Exact Solution for Piston like Leak-off of Compressible Fluids. Journal of Energy Resources Technology 1983December, 106

42. A Settari, A New General Model of Fluid Loss in Hydraulic Fracturing, SPE 11625, August 1985 1985 491 501

43. S. L Crouch, A. M Starfield, Boundary Element Methods in Solid Mechanics. 1st ed. London: George Allen & Unwin Ltd, 1983

44. J. E Olson, Predicting fracture swarms: the influence of subcritical crack growth and the crack-tip process zone on joints spacing rock. In: Cosbrove JW, Engeder T (eds) The initiation, propagation and arrest of joints and other fractures. Geological Soc. Special Publications, London, 2004L 231 73 87

45. M. G Mack, N. R Warpinski, Mechanics of Hydraulic Fracturing. In: Reservoir Stimulation, Third Edition. Editors: Economides MJ and Nolte KG, 2000

Chapter 10

HYDRAULIC CONDUCTIVITY OF SEMI-QUASI STABLE SOILS: EFFECTS OF PARTICULATE MOBILITY

Oagile Dikinya

Department of Environmental Science, University of Botswana, Botswana

INTRODUCTION

Particulate mobility within the intra-soil aggregates and soil pores has profound effects on the stability of soils and consequently hydraulic conductivity. Soil structure or aggregation essentially describes the way the soil constituents (sand, silt, clay, organic matter) are arranged and the size and shape of pores between them (Geeves et al. 1996). On the basis of size, soil aggregates can be distinguished as macro-aggregates (> 250 μm) and microaggregates (< 250 μm) and with further breakdown, they can release finer particles (< 20 μm, Oades and Waters 1991). However, aggregates < 2 μ m (predominately clay floccules) are held together by forces derived from the interaction of clay particles. The loss of macroporosity is attributable to detachment and subsequent transport of aggregates (Roth et al. (1991; Sutherland et al. 1996) and further dis-aggregation can affect soil water retention (Neufeldt et al. 1999). Aggregates with low stability fracture easily and the breakdown into smaller sizes is a major cause reduced permeability through pore clogging as a result of loss of aggregation and porosity of surface soils. Once a crust is formed the water and solute transport properties change, for example increased rate of water infiltration with removal of crust by application of gypsum (Ramirez et al.,1999). This surface crusts or sealing is often attributable to slaking and dispersion of clay minerals.

Dispersion is most commonly associated with sodicity and is highly sensitive to both the exchangeable sodium percentage (ESP) and total electrolyte concentration. Soils with a high ESP are more susceptible to dispersion due to its greater ability to develop diffuse double layers. The exchangeable sodium percentage (ESP) can be defined as:

$$ESP = \frac{(100 * ExchangeableNa)}{CEC} \tag{1}$$

where, CEC is the cation exchange capacity.

Similarly, Sodium adsorption ratio (SAR) can also be used to define sodicity, particularly when using soil solutions.

$$SAR = \frac{\left[Na^+\right]}{\left(0.5\left(\left[Ca^{2+}\right]+\left[Mg^{2+}\right]\right)\right)^{0.5}}$$

(2)

Where, all [concentrations are expressed as mmol(+)/L].

In general, high levels of SAR decrease the stability of soil structures and microstructures usually become more unstable, deflocculated and dislodged as the ESP increases with more production of fine or mobile particles. For example, in particulate facilitated transport, mobile colloids must be present in large concentrations and must be transported over significant distances (Kretzschmar et al. 1999). In most environments, mobilization is favoured by high pH, high SAR and low ionic strength resulting in severe decreases in permeability. Permeability or hydraulic conductivity is estimated using Darcy's law for one-dimensional vertical flow (Klute and Dirksen 1986):

$$q = K * \partial H / \partial z$$

(3)

where q is the flux density, K is the hydraulic conductivity, $\partial H / \partial z$) is gradient of hydraulic head H ,and z is the gravitational head.

Despite the considerable amount of research carried out, the processes by which the structural breakdown and pore clogging by particulate transport occur in soils are far from satisfactorily understood and also limited by suitable modelling procedures. For instance McDowel-Boyer et al.(1986) and Harvey and Garabedian (1991) by applying filtration theory, attributed poor estimates of transport with natural porous media to; (i) the wider pore size and particle size distributions, complex pore geometry and rough matrix surface or these materials and (ii) a wide particle size distribution mobile surface colloids. Using a cake filtration model, Mays and Hunt (2005) suggested that the increased head loss results from the formation of deposits with a decline in hydraulic conductivity. Reduced flow capacity of the soil matrix can be induced by the invasion and geochemical transformation due to plugged pore channels (Song and Elimelech, 1995) thus minimising erosion of base soils with reduction in filter permeability (Lee et al., 2002).

Thus research was undertaken in an attempt to throw light on the dynamics of the processes of mobilisation during the flow of NaCl solutions through two soil materials of markedly different structural cohesiveness. The experiments were carried using saturated soil columns to determine hydraulic conductivity and particle size distribution as the criterion for the assessment of the dynamic flow processes.

MATERIALS AND METHODS

Soil Samples and Their Characteristics

Two different soil samples were used as porous media; sample A and B with distinct physico-chemical characteristics (Table 1) analysed using standard methods (Klute 1986; Klute 1986b). Sample A provides a classic example of a relatively fragile agricultural soil which is highly susceptible to structural breakdown and permeability problems while sample B is a completely disrupted and reconstituted material (sand and clay) and presents similar structural problems.

Table 1. Physico-chemical characteristics of soil samples.

Soil Property	Sample A	Sample B
Sand (%)	83.3	89.0
Silt (%)	6.6	1.9
Clay (%)	10.1	9.1
Texture	Sandy loam	Loamy sand
Electrical conductivity ($\mu S/cm$)	42	380
pH(water)	6.1	8.7
CEC ($mmol_c/g$)	3.4	8.1
Organic Carbon (%)	0.03	0.55
Clay mineralogy	Kaolinite,vermiculite	Kaolinte, smectite
Bulk density (g/cm^3)	1.61	1.58

CEC-cation exchange capacity

Particle Size Distribution Measurements

The particle size distributions (Table 2) for the various sized fractions of these porous media were determined using the sedimentation or grain size mechanical wet sieving method (Day 1965) for the particle size range from 45-2000 μm while a laser light scattering technique (Mastersizer Microplus Ver.2.18, c/o Malvern Instruments Ltd, 1995) was used for the range < 45 μm. A mixture of air dried soil (<2 mm) with water was boiled and 6 % hydrogen peroxide solution was added to remove organic matter followed by a calgon/NaOH mixture to disperse the soil. Suspensions were collected for various fraction sizes according to the sedimentation theory or Stokes's law (Stokes, 1891). For the sand fractions suspensions were separated using sieves of various diameters (1000, 500, 250, 125, 45 μm). Suspensions passing through 45 μm together with effluents from subsequent leaching experiments were analyzed for particle size distribution using a Malvern Mastersizer analyser following leaching experiments. Suspensions were dispersed in the Mastersizer's ultrasonic bath unit (equipped with a small angle light scattering apparatus, Helium-neon laser; λ = of 633 nm, as a light source) for about 25 minutes. Suspension concentrations were adjusted until an obstruction of a least 0.2 % is reached for best results using refractive indices of 1.59 and 1.33, respectively for clay and deionised water with a particle density of 2.6 g/cm^3. Both values assumed to be representative for the soil material used for analysis.

Table 2. Particle size fractions for samples A and B.

Size fractions (μm)	Sample A	Sample B
	% particle fraction (g/g)	
1000 – 2000 μm	12.37	1.70
500 – 1000 μm	26.78	12.81
250 – 500 μm	21.30	57.46
125 – 250 μm	15.92	13.67
45 – 125 μm	2.41	2.40
< 45 μm	4.52	1.00

Hydraulic Conductivity Measurements

Constant head Hydraulic Conductivity Measurements

A glass Marriotte bottle, filled with deionised water was used to maintain a constant head, with ambient temperature during the experiment varying from 20 to 22 °C. The flow was from the bottom to the top of the columns to prevent air entry. The hydraulic head across the soil samples (ΔH) was kept at an average of 30 cm and 40 cm for samples A and B, respectively. ΔH was measured as equivalent to the vertical distance from upper water level (on the Marriotte bottle) to the bottom at the leachate collection point. The flow was continuous during leaching in order to maintain saturation throughout the experiments. In the case of flow interruptions, columns were kept saturated by tightly closing the system to avoid air entry. Hydraulic condtivity (HC) was initially measured using the deionised water followed by 1mmol NaCL to subsequently measure pore size distribution before and after leaching with 1mmol NaCL. Any turbid percolate observed during the experiment was collected and stored in plastic bottles for further analysis of effluent particle size distribution using the Malvern Mastersizer technique, and for mineralogical characterization. Using a column and constant head experimental set up, equation 3 can be re-written as $K = (4VL)/(\pi d^2 \Delta t \Delta H)$:where V (cm^3) is the volume of water collected during time interval Δt (minutes) , L (cm) is the length of column soil sample, d (cm) is the inner diameter of the column and ΔH (cm) is the change in hydraulic head across the soil sample.

Pore-size distributions of the soil columns were estimated from water-retention versus water-potential curves determined using sintered glass funnels and the Haines method over the low pressure ranges (measured at 5 cm suction interval) and pressure-plate, ceramicmembrane apparatus for the higher pressure range (10 to 800 KPa). The column samples were pre-wetted with 0.1M $CaCl_2$ solution to avoid swelling and subsequently leached with deionised water followed by 1 mmol/L NaCL solution adjusted at pH 8.5. Since the soils were predominately kaolinitic in nature and therefore have low swelling potential. Similarly, the water-retention data was transformed to pore-size distribution using the form of Kelvin equation:

$$\Delta P = (2\sigma / rp) \qquad (4)$$

where, ΔP is the pressure difference (Pa) across an air-water interface, σ is the surface tension of water(Jm-2) and rp is the radius of a circular capillary tube (m).

Pressure Leaching Experiments

Pressure leaching simulations and/or experiments were carried out using soil columns as shown schematically in Figure 1. Soil samples (<2 mm) were initially mixed with 0.9–1 mm acid-washed sand to provide a rigid skeletal structure to facilitate flow and colloidal mobility. The proportions of mixtures were 50% soil–50% sand and 30% soil–70% sand, for samples B and A, respectively. The mixtures were uniformly wet packed into acrylic soil columns (400 mm long, 25 mm diameter, 2 mm wall thickness, cross-sectional area 4.91 cm2, volume 196 cm3) and saturated with the desired electrolyte concentrations. Holes were bored at various column depths (Xi), for horizontal insertion of pressures sensors (Pi) at 50 mm (X1) and 250 mm (X2), along the column from the inlet (Fig. 1). Miniature pressure transducer with tensiometers T5–7/5 model (UMS, 2000) each with 1.5 m cable length and 5 cm shaft length and 20 mm diameter were connected to the system for online pressure monitoring and measurements of saturated hydraulic conductivity with time.

A peristalitic pump (designed with a system maximum discharge pressure of 300 kPa at 0 – 40° C), was used to pump solutions of ionic strengths of 1 mmol/L NaCl into the soil columns in a vertical upward flow. During leaching, pressures were measured. For continuous online pressure measurements, an Agilent 34970A model data logger-PC system (Agilent Technologies 2003) was used as a data acquisition system. To ensure equilibration of the system and to enable particle-hydrodynamic settling, the columns were saturated with 1 mmol/L NaCl solution overnight (for at least 12 hours). Prior to experimental runs, a steady state was attained by measuring outflow rates until a constant value was achieved. Hydraulic conductivity was then estimated from equation 3 (Darcian flow under hydrodynamic dispersion). Further, to simulate or quantify the effect of ionic strength, the columns were also leached with 10, 2.5 and 1 mmol/L NaCL solution to measure the extent of hydraulic conductivity decrease in response to the electrical conductivity.

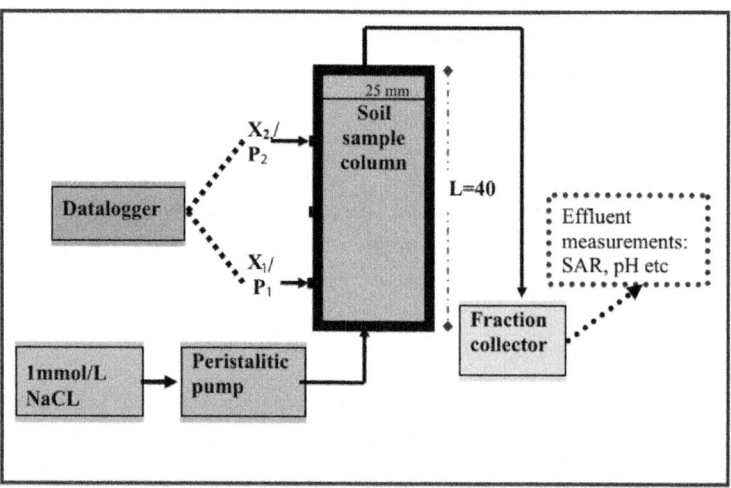

Figure 1. Schematic diagram for pressure leaching experiment for hydraulic conductivity measurements ($X_{1(inlet)}$ = 50 mm and $X_{2(outlet)}$ = 250 mm from bottom of column and P1 and P2 are respective pressures points).

RESULTS AND DISCUSSION

Hydraulic Properties Associated With Dispersion of Aggregates

The pore-size distributions were obtained from water retention data (θ vs h) in Figs. 2a and 3a for sample A and B, respectively; where θ is volumetric water content and h is suction or pressure head. The pore radius r of samples were generating by differentiating $d\theta/dh$ from the measured water contents and pressure head or suction measurements. The pore-size distributions for the columns packed with samples are shown in Figs. 2b and 3b for sample A and B, respectively. The pore size distribution for the sample A soil columns were broader with the peak occurring at approximately 8 μm while those for the sample B were much narrower with a peak at approximately 12 μm.

The small shift in the narrow distribution for Sample A following leaching would be consistent with small dispersion and relocation of particles (see Figs 4 and 5). The broader distribution for the Sample B residue became substantially narrower following leaching (Figs 2 and 3). This change would explain the dramatic decrease of hydraulic conductivity by 4 orders of magnitude (See section 3.2). Similar observations were made by Leij et al. (2002) who observed that over the total number of pores decrease, the mean pore size decreased from an initial 49.4 to 28 ┤ m after disturbance. This dispersion-dependent pore size distribution changes in samples were manifested by decrease of porosities: 0.370 to 0.354 cm³/cm³ for sample A and 0.431 to 0.417 cm3/cm3 for sample B following leaching with 1 mmol/L NaCl. i.e a decrease of porosity by 9% and 3% for sample A and B, respectively. This accounted for the subsequent decreases of actual hydraulic conductivity from 6.1 to 0.02 cm/h and from 1.5 to 0.14 cm/h for sample A and B, respectively (see section 3.2)

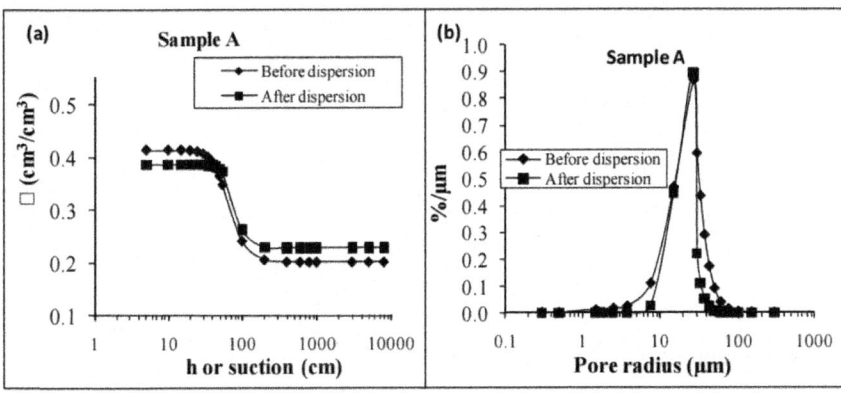

Figure 2. Soil water retention (a) and pore-size distribution (b) for sample A.

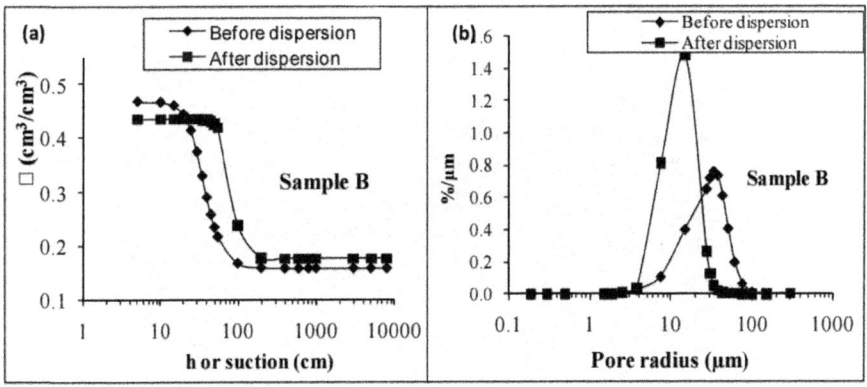

Figure 3. Soil water retention (a) and pore-size distribution (b) for sample B.

Similarly, using measured particle size distribution (Fig 4), soil structure was modelled at the pore scale (Dikinya et al., 2007) to explain the different response of the two samples to the experimental conditions (see Fig 5). The size of the pores was determined as a function of deposited clay particles. The modal pore size of sample B as indicated by the constant water retention curve was 45 μm and was not affected by the leaching process. In the case of the sample B, the mode changed from 75 to 45 μm. This reduction of pore size corresponds to an increase of capillary forces that is related to the measured shift of the water retention curve of the two samples.

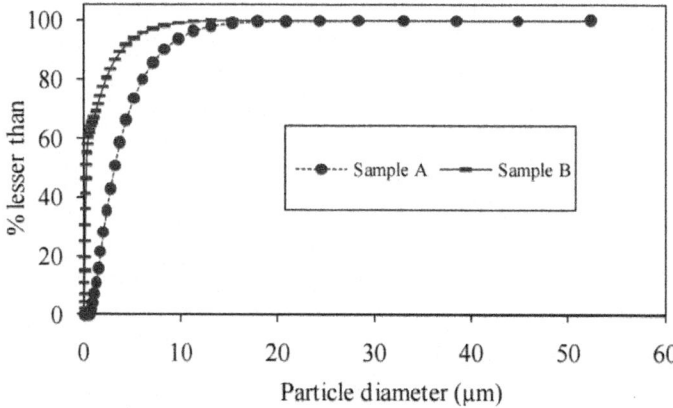

Figure 4. Particle size distribution of effluents.

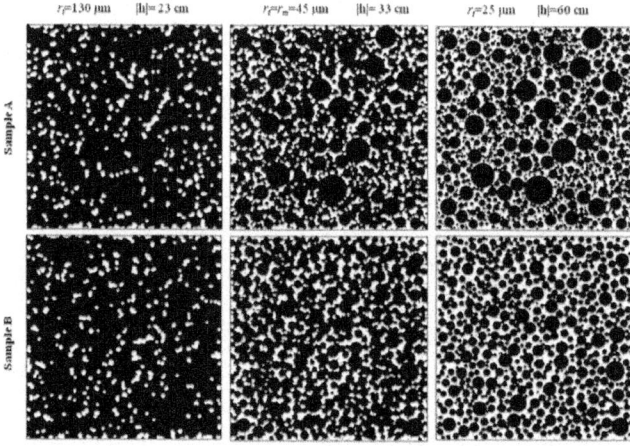

Figure 5. Size-dependent pore connectivity for soil samples A and B. All particles and pores smaller than the threshold value r_t, are shown in black and pores of radius $r \geq r_t$ in white. Source: [Dikinya et al. 2007].

Comparison of the modelled pore size distribution with the measured water retention curves for the four cases; (i) Sample B before the release process, (ii) Sample A before the release, (iii) Sample B after the re-deposition and (iv) Sample A after the re-deposition, have revealed that measured water retention curves were very similar for the cases i), iii) and iv). The air entry values of these three curves were 2-3 times the value observed for sample B before the release process (ii). The same phenomenon was simulated with the pore model: the mode of the modelled pore size distribution was similar (about 45µm) for (i), (iii) and (iv) and about 75 to 95 µm for ii). The ratio of the capillary forces associated with these values of the modes is somehow equal to 1.67 to 2.11 and is similar to the measured ratio of the air entry values.

Hydraulic Conductivity under Constant Head Using Mariotte Bottle Device

The hydraulic conductivity (HC) was found to progressively decrease with time. This decrease was attributed to decreases of pore radii associated with increased instability of soil intra-aggregates during leaching. Sample B was substantially more prone to structural disintegration than sample A with actual hydraulic conductivity decreases from 6.1 to 0.02 cm/h and from 1.5 to 0.14 cm/h, respectively as manifested by a relative marked measured shift of the water retention curve. In sample B there was appreciable decrease of the HC to 8.5 % of the initial value. This was attributed to localized pore clogging (similar to a surface seal) affecting hydraulic conductivity, but not the microscopically measured pore size distribution or water retention. This decrease could be explained with the modelled destruction of large pores due to particle re-deposition.

Hydraulic Conductivity Associated With Particle Mobility and Pressure Build-Up

The dynamics of the process of structural disintegration in soil columns were further evaluated by simultaneously measuring changes in pressure gradients along the columns and sodium adsorption ratio (SAR) during pressure leaching with solutions of 10, 2.5, 1 mmol/L NaCl. Steady increases in pressure gradient ($\Delta P/\Delta L$) and corresponding decreases in RHC with time were observed for both soils and follow similar trends at all column depths indicating continuous particle accumulation in filter pores. The most severe increases in $\Delta P/\Delta L$ and decreases in RHC always occurred near the inlet to the columns and the decline gradually decreased along the column. An increase of $\Delta P/\Delta L$ and decrease in RHC with decreasing ionic strengths was also observed for both soils. The decreases in RHC and increases in $\Delta P/\Delta L$ were clearly influenced by the size as well as the concentration of migrating particles in the porous medium. The finer mobile particles in the mining residue were clearly more readily self-filtered at the lower electrolyte concentration than the larger sample B particles, producing more rapid increases in $\Delta P/\Delta L$ and decreases in RHC. This more effective particulate movement and more rapid plugging is undoubtedly due to increased development of the diffuse double layer, swelling and dispersion within the soil matrix at these concentrations.

A more dramatic decrease of saturated hydraulic conductivity was noticeable for the less cohesive sample A than for the aggregated sample B. Due to differences in the clay mineralogy and its treatment history involving greater disruption, sample A is far more sensitive to particle mobilisation and pore clogging than sample B. Sample A is therefore more likely to encounter constriction and retention thus allowing smaller sized particles to become entrapped behind the coarser fraction, forming a filtration zone. The more clogging occurs the greater is the decrease of hydraulic conductivity and the smaller is the amount of particles being released. This particularly enhanced at the lower electrolyte concentrations. Further, RHC of both soil samples has been shown to decrease with time and with increasing SAR with the reduction being substantially greater for the less cohesive sample B. This was probably due to dilution of high-sodicity soil irrigation water that can cause induce swelling, aggregate slaking and particle clay dispersion (Bagarello et al. 2006).

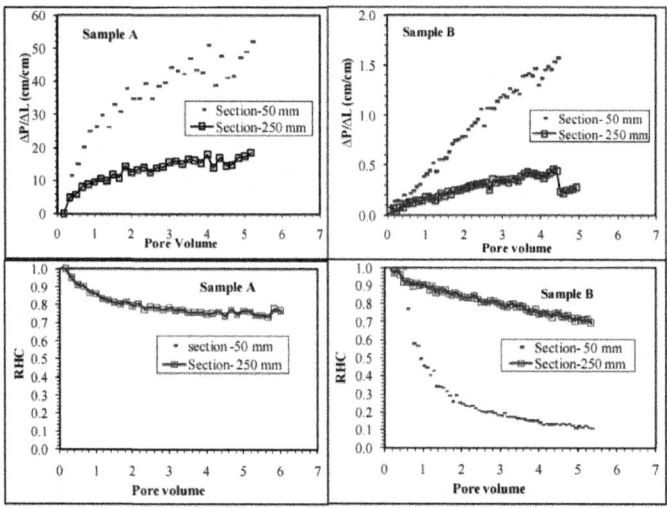

Figure 6. Pressure build up (upper curves) and hydraulic conductivity (lower curves) attributable to structural disintegration and particle migration during leaching experiments.

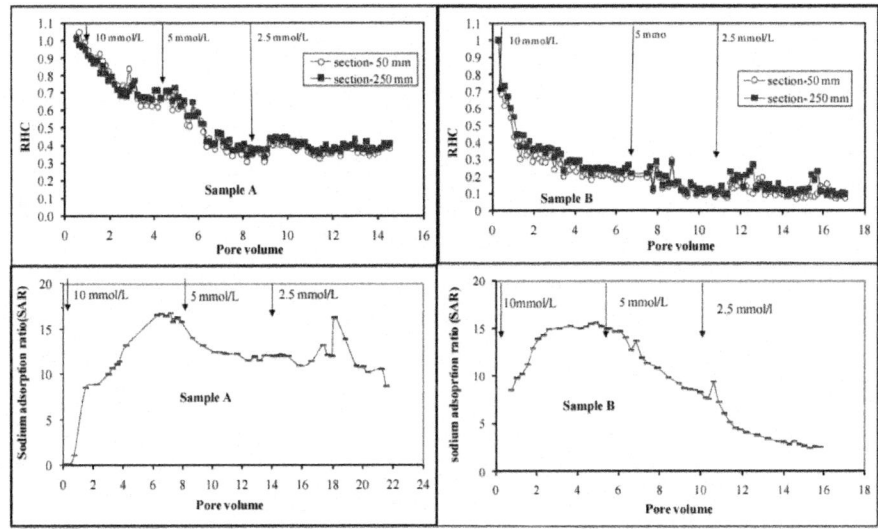

Figure 7. Effect of ionic strength on relative hydraulic conductivity (RHC) and its associated sodium adsorption ratio (SAR) during leaching experiments.

CONCLUSIONS

The dynamics of the process of structural disintegration in soil columns were evaluated by simultaneously measuring changes in pressure gradients along the columns and concentrations during pressure leaching. The decreases in RHC and increases in $\Delta P/\Delta L$ were clearly influenced by the size as well as the concentration of migrating particles in the porous medium. The finer mobile particles in Sample B were clearly more readily selffiltered at the lower electrolyte concentration than the larger sample A, producing more rapid increases in $\Delta P/\Delta L$ and decreases in RHC. This more rapid decrease in RHC particularly at the lowest concentration (1 mmol/L) was consistent with measured SAR. The effects of clay mineralogy are evident with the kaolinite-smectite sample B having a more marked decrease in RHC with increasing ionic strength compared with the kaolinitic sample A clays.

REFERENCES

1. Bagarello V, Iovino M, PalazzoloE, Panno M, Reynolds WD (2006) Field and laboratory approaches for determining sodicity effects on saturated soil hydraulic conductivity. Geoderma 130, 1–13.

2. Day PR (1965) Particle fractionation and particle-size analysis. In 'Methods of soil analysis part I'. Agronomy 9 (Ed. CA Black) pp 545–

567, (American. Society of Agronomy Madison, Wisconsin).

3.   Dikinya O, Lehmann P, Hinz C and Aylmore LG (2007) Using a pore scale model to quantify the effect of particle re-arrangement on pore structure and hydraulic properties. Hydrological Processes, 21, 989–997.

4.   Greene RS, Nettleton WD, Chartres CJ, Leys JF, Cunningham RB (1998) Runoff and micromorphological properties of a grazed haplargid near Cobar, NSW, Australia. Australian Journal of Soil Research 36, 87–108.

5.   Geeves G, Cresswell H, Murphy B, Chartres C (1996) Productivity and sustainability from managing soil structure. Extension brochure. (CSIRO Division of Soils and NSW Department of Land and water conversation: Canberra)

6.   Harvey RW, Garabedian SP (1991) Use of colloid filtration theory in modelling movement of bacteria through a contaminated sandy aquifer. Environmental Science Technolology 25, 178–185

7.   Klute A (1986b) Water retention: Laboratory methods. In "Methods of soil analysis. Part 1– Physical and Mineralogical Methods. Second Edition". Agronomy Monograph No 9: American Society of Agronomy/ Soil Science Society of America (Ed. A Klute), Madison, Wisconsin.

8.   Klute A (1986) Methods of soil analysis. Part 1–Physical and Mineralogical Methods. Second Edition". Agronomy Monograph No 9: American Society of Agronomy/ Soil Science Society of America (Ed. A Klute), Madison, Wisconsin.

9.   Kretzschmar R., Borkovec M, Grolimund D, Elimelech M (1999) Mobile Subsurface Colloids and their role in contaminant transport. Advances in Agronomy 65, 1 – 96.

10.   Leij FL, Ghezzehei TA, Or D (2002) Analytical models for soil pore-size distribution after tillage. Soil Science Society of America Journal 66, 1104–1114.

11.   Lee IM, Park YJ, Reddi LN (2002) Particle transport characteristics and filtration of granitic residual soils from the Korean peninsula. Canadian Geotechnical Journal 39, 472–482.

12.   Lehmann P, Wyss P, Flisch A, Lehmann E, Vontobel P, Kaester A, Beckmann F, Frey O, Gygi A, Flühler H (2006) Mapping the structure of porous media using tomography with X-ray, thermal neutrons and synchrotron radiation, Vadoze Zone Journal 5, 80– 97; doi:10.2136.

13.   Malvern Instruments Ltd (1995). "Malven Mastersizer Microplus version ver 2.18". (Engigma Business Park: Gronwood Road, UK).

14.   Neufeldt H, Ayarza MA, Resck DVS, Zech W (1999) Distribution of

water-stable aggregates and aggregating agents in Cerrado. Geoderma 93, 85–99.

15. Roth CH, de Castro Filho C, de Mediros GB (1991). Analise de fatores fiscos e quimicos relacionados com a agregacao de um latossolo xoxo distrofico. R Bras. Ci. Solo 15, 241–248.

16. Song L, Elimelech M (1995) Particle deposition onto a permeable surface in laminar flow. Journal of colloid and interface science 173, 165–180. Stokes GG (1891) Mathematical and Physical Papers III, Cambridge Press.

17. Sutherland RA, Watung RL, El-Swaify (1996) Splash transport of organic carbon and associated concentrations and mass enrichment ratios for Oxisol, Hawaii. Earth surface processes landforms 21, 1145–1162.

Chapter 11

ELECTROKINETIC TECHNIQUES FOR THE DETERMINATION OF HYDRAULIC CONDUCTIVITY

Laurence Jouniaux

Institut de Physique du Globe de Strasbourg, Université de Strasbourg, Strasbourg France

INTRODUCTION

In a porous medium the fluid flux and the electric current density are coupled, so that the streaming potentials are generated by fluids moving through porous media and fractures. These electrokinetic phenomena are induced by the relative motion between the fluid and the rock because of the presence of ions within water. Both steady-state and transient fluid flow can induce electrokinetics phenomena. It has been proposed to use this electrokinetic coupling to detect preferential flow paths, to detect faults and contrast in permeabilities within the crust, and to deduce hydraulic conductivity. This chapter proposes a comprehensive review of the electrokinetic coupling in rocks and sediments and a comprehensive review of the different approaches to deduce hydraulic properties in various contexts.

Electrical methods are sensitive to the fluid content because of the relative high conductivity of water compared to the one of the rock matrix. The electrical resistivity can be related to the permeability and to the deformation, in full-saturated or in partially-saturated conditions (Doussan & Ruy, 2009; Henry et al., 2003; Jouniaux et al., 1994; 2006). The electrokinetic phenomena are induced by the relative motion between the fluid and the rock matrix. In a porous medium the electric current density, linked to the ions within the fluid, is coupled to the fluid flow (Overbeek, 1952) so that the streaming potentials are generated by fluids moving through porous media (Jouniaux et al., 2009). The classical interpretation of the self-potential (SP) observations is that they originate from electrokinetic effect as water flows through aquifer or fractures. Therefore some formula have been proposed to predict the permeability of porous medium or fault using the electrokinetic properties. The SP method consists in measuring the natural electric field on the Earth's surface. Usually the electric field is measured by a high-input impedance multimeter, using impolarizable electrodes (Petiau, 2000; Petiau & Dupis, 1980) and its interpretation needs filtering techniques (Moreau et al., 1996). Moreover, for long-term observations the monitoring of the magnetic field is also needed for a good interpretation (Perrier et al., 1997). Some studies have proposed to use SP observations to infer water-table variations, to estimate hydraulic properties (Glover & Walker, 2009), and to deduce where to make a borehole for water-catchment. These studies involve surface or borehole measurements (Aubert & Atangana, 1996; Fagerlund & Heison, 2003; Finizola et al., 2003; Perrier et al., 1998; Pinettes et al., 2002), some of them have monitored self-potentials during hydraulic tests in boreholes (Darnet et al., 2006; Darnet & Marquis, 2004; Ishido et al., 1983; Maineult et al., 2008). Direct models (Ishido & Pritchett, 1999; Jouniaux et al., 1999; Sheffer & Oldenburg, 2007) and inverse problems (El-kaliouby & Al-Garni, 2009; Fernandez-Martinez et al., 2010; Gibert & Pessel, 2001; Gibert & Sailhac, 2008; Minsley et al., 2007; Naudet et al., 2008; Sailhac et al., 2004; Saracco et al., 2004) have been developed to locate the source of self-potential. Because of similarity between the electrical potential with pressure behavior, it has been proposed also to use SP measurements as an electrical flow-meter (Pezard et al., 2009). However, inferring a firm link between SP intensity and water flux is still difficult. Recent modeling has shown that SP observations could detect at distance the propagation of a water front in a reservoir (Saunders et al., 2008).

We distinguish 1) The steady-state and passive observations which consist in measuring the electrical self-potential (SP). 2) The transient and active observations which consist in measuring the electrical potential induced by the propagation of a seismic wave. These observations are called seismo-electric conversion. The reverse can also be observed: the detection of a seismic wave induced by injection of electrical current and is called electro-seismic conversion.

STREAMING POTENTIAL COEFFICIENT IN ROCKS AND SEDIMENTS

Theoretical Background

The fluid flow in porous media or in fractures can induce electrokinetic effect because of the presence of ions within the fluid which can induce electric currents when water flows. The general equation coupling the different flows is,

$$J_i = \sum_{j=1}^{N} \mathcal{L}_{ij} X_j$$

$$(1)$$

which links the forces X_j to the macroscopic fluxes J_i, through transport coupling coefficients L_{ij} (Onsager, 1931).

When dealing with the coupling between the hydraulic flow and the electric flow, assuming a constant temperature, and no concentration gradients, the electric current density Je [A.m^{-2}] and the flow of fluid J_f [m.s^{-1}] can be written as the following coupled equation:

$$J_e = -\sigma_0 \nabla V - \mathcal{L}_{ek} \nabla P.$$

$$(2)$$

$$J_f = -\mathcal{L}_{ek} \nabla V - \frac{k_0}{\eta_f} \nabla P.$$

$$(3)$$

where P is the pressure that drives the flow [Pa], V is the electrical potential [V], σ_0 is the bulk electrical conductivity [S.m^{-1}], k_0 the bulk permeability [m^2], η_f the dynamic viscosity of the fluid [Pa.s], L_{ek} the electrokinetic coupling [A Pa^{-1} m^{-1}]. Thus the first term in equation 2 is the Ohm's law and the second term in equation 3 is the Darcy's law. The coupling coefficients must satisfy the Onsager's reciprocal relation in the steady state: the coupling coefficient is therefore the same in equation 2 and equation 3. This reciprocity has been verified on porous materials (Auriault & Strzelecki, 1981; Miller, 1960) and on natural materials (Beddiar et al., 2002).

The streaming potential coefficient C_{s0} [V.Pa^{-1}] is defined when the electric current density J_e is zero, leading to

$$\frac{\Delta V}{\Delta P} = -\frac{\mathcal{L}_{ek}}{\sigma_0} = C_{s0}$$

$$(4)$$

This coefficient can be measured by applying a driving pore pressure ΔP to a porous medium and by detecting the induced electric potential difference ΔV. The driving pore pressure induces a streaming current (second term in eq. 2) which is balanced by the conduction current (first term in eq.2) which leads to the electric potential difference ΔV that can be measured. We detail here what we know about this streaming potential coefficient (SPC) on sands and rocks because we will see that it can be used with the electro-osmosis coefficient to deduce the permeability. In the case of a unidirectional flow through a cylindrical saturated porous capillary, this coefficient can be expressed as (Jouniaux et al., 2000; Jouniaux & Pozzi, 1995b):

$$C_{s0} = \frac{\epsilon_f \zeta}{\eta_f \sigma_{eff}}$$

$$(5)$$

with the fluid electrical permittivity ε_f [F.m-1], the effective electrical conductivity σ_{eff} [S.m^{-1}] defined as $\sigma_{eff} = F\sigma_0$ with F the formation factor and $\sigma0$

the rock conductivity which can include a surface conductivity. The potential ζ [V] is the zeta potential described as the electrical potential inside the EDL at the slipping plane or shear plane (i.e., the potential within the double-layer at the zero-velocity surface). Minerals forming the rock develop an electric double-layer when in contact with an electrolyte, usually resulting from a negatively charged mineral surface. An electric field is created perpendicular to the surface of the mineral which attracts counterions (usually cations) and repulses anions in the vicinity of the pore matrix interface. The electric double layer (Fig. 1) is made up of the Stern layer, where cations are adsorbed on the surface, and the Gouy diffuse layer, where the number of counterions exceeds the number of anions (Adamson, 1976; Davis et al., 1978; Hunter, 1981). The streaming current is due to the motion of the diffuse layer induced by a fluid pressure difference along the interface. This streaming current is then balanced by the conduction current, leading to the streaming potential. When the surface conductivity can be neglected compared to the fluid conductivity $F\sigma_0 = \sigma_f$ and the streaming coefficient is described by the well-known Helmholtz-Smoluchowski equation (Dukhin & Derjaguin, 1974):

$$C_{s0} = \frac{\epsilon_f \zeta}{\eta_f \sigma_f}$$

(6)

The assumptions are a laminar fluid flow and identical hydraulic and electric tortuosity. The influencing parameters on this streaming potential coefficient are therefore the dielectric constant of the fluid, the viscosity of the fluid, the fluid conductivity and the zeta potential, itself depending on rock, fluid composition, and pH (Guichet et al., 2006; Ishido & Mizutani, 1981; Jaafar et al., 2009; Jouniaux et al., 2000; Jouniaux & Pozzi, 1995a; Lorne et al., 1999a; Vinogradov et al., 2010). There exists a pH for which the zeta potential is zero: this is the isoelectric point and pH is called pH_{IEP} (Davis & Kent, 1990; Sposito, 1989). At a given pH the most influencing parameter is the fluid conductivity (Fig.2). When collecting data from literature on sands and sandstones we can propose that $C_{s0} = -1.2 \times 10^{-8} \sigma_f^{-1}$ which leads to a zeta potential equal to -17 mV assuming eq. 6 and that zeta potential and dielectric constant do not depend on fluid conductivity. These assumptions are not exact, but the value of zeta is needed for numerous modellings which usually assume the dielectric constant not dependent on the fluid conductivity. Therefore an average value of -17 mV for such modellings is rather exact, at least for medium with no clay nor calcite. Another formula is often used (Pride & Morgan, 1991) based on quartz minerals rather than on sands and sandstones, which may be less appropriate for field applications. When the medium is not fully saturated Perrier & Morat (2000) suggested a model in which the streaming potential coefficient is dependent on a relative permeability model k_r.

$$C(S_w) = C_{s0} \frac{k_r}{S_w{}^n} \qquad (7)$$

assuming that the relative electrical conductivity is equal to S_w. The parameter n is the Archie saturation exponent (Archie, 1942). This exponent has been observed to be about 2 for consolidated rocks and in the range $1.3 < n < 2$ for coarse-texture sand (Guichet et al., 2003; Lesmes & Friedman, 2005; Schön, 1996). Note that the use of Archie's law is valid in the absence of surface electrical conductivity. Recently Allègre et al. (2010) (and Allègre et al. (2011)) proposed original streaming potential measurements performed during a drainage experiment and measured the first continuous recordings of the streaming potential coefficient as a function of water saturation. These authors observed that the streaming potential coefficient exhibits two different behaviours as the water saturation decreases. Values of C_{s0} first increase for decreasing saturation in the range $0.55 - 0.8 < S_w < 1$, and then decrease from $S_w = 0.55 - 0.8$ to residual water saturation. This behaviour was never reported before and still needs further interpretation. Jackson (2010) used a bundle capillary model to compute the streaming potential coeffic cent as a function of water-saturation. He showed that the behaviour of the SPC depends on the capillary size distribution, the wetting behaviour of the capillaries, and whether we invoke the thin or thick electrical double layer assumption. Depending upon the chosen value of the saturation exponent and the irreductible water-saturation, the relative SPC may increase at partial saturation, before decreasing to zero at the irreductible saturation. Up to now permeability predictions using electrokinetic techniques use theoretical developments in full saturated conditions.

Similarly the electro-osmosis coefficient is defined when the flow of fluid J_f is zero, leading to

$$\frac{\Delta P}{\Delta V} = -\frac{\mathcal{L}_{ek}\eta}{k_0} = C_{e0} \qquad (8)$$

This coefficient can be measured by applying an electric potential difference ΔV and by detecting the induced electro-osmotic flow [m.s^{-1}] corresponding to the first term of equation 3, by controlling the hydraulic gradient, usually maintaining identical water heads. Assuming the Helmholtz-Smoluchowski equation (eq. 6) the electro-osmosis coefficient can be written as:

$$C_{e0} = \frac{\epsilon\zeta}{k_0 F} \qquad (9)$$

and then depends also on pH (Beddiar et al., 2005) through the zeta potential. Since the permeability and the formation factor are not independent, but can be related by $k_0 = CR^2/F$ (Paterson, 1983) with C a geometrical constant usually in the range 0.3-0.5 and R the hydraulic radius, the electro-osmosis coefficient can be written as:

$$C_{e0} = \frac{\epsilon\zeta}{CR^2} \qquad (10)$$

As we can see from this section, the streaming potential coefficient and the electro-osmosis coefficient are directly proportional to the zeta potential, which cannot be directly measured and which is difficult to model at a rock-water interface. Therfore the zeta potential is usually deduced from streaming potential measurements. Moreover the streaming potential coefficient is inversely proportional to the fluid conductivity, whereas the electro-osmosis coefficient is inversely proportional to the hydraulic radius.

Permeability Prediction

These electrokinetic properties have been used to predict the permeability. Li et al. (1995) defined an electrokinetic permeability k_e by the following relation:

$$k_e = \eta\sigma_r\frac{C_{s0}}{C_{e0}} \qquad (11)$$

with σ_r the rock conductivity (measured when J_f is zero).

These authors verified on 12 samples of sandstones, limestones and fused glass beads that the electrokinetic permeability k_e successfully predicts the rock permeability k_r (measured when Je is zero) over a range of about four decades from 10^{-15} to 10^{-11} m^2 . Pengra et al. (1999) verified also this relation on eight samples of sandstone and limestone, and four fused glass beads, in the permeability range 10^{-15} to 10^{-11} m^2 (Fig. 3). This approach has been used to propose the permeability measurement (Wong, 1995) within boreholes (Fig. 4). The advantage was that we only needed to apply or to measure the pressure and the electric field.

A simplest way to measure the permeability in borehole, was performed using only the streaming potential coefficient (eq. 4). Although this coefficient does not depend directly on permeability, Hunt & Worthington (2000) showed that the borehole streaming potential response could detect fractures and cracks. A pressure pulse is generated by a nylon block which displaces water as it moves upwards (Fig. 5). This mechanical system avoids spurious electrical noise induced by electro-mechanical systems. The electrode response is normalized to the peak pressure recorded by the hydrophone. The authors showed that the maximum electrical signal was clearly associated with the highest fracture density and the widest aperture (few cm). The recorded amplitudes were in the range 4×10^{-7} to 1.5×10^{-6} V/Pa. It was proposed that the fluid flow in the cracks causing the streaming potential was predominantly caused by the seismic wave within the rock that distorts crack aperture as it passes, rather than by the source directly forcing fluid into cracks. In this case the permeability dependence of the streaming potential coefficient may be linked to the indirect effect of surface conductivity which may not be negligeable: the effective conductivity can decrease with increasing permeability, leading to an increase in the streaming potential coefficient (eq. 5) (Jouniaux & Pozzi, 1995a).

Recently, Glover et al. (2006) proposed a new prediction for the permeability by comparing an electrical model derived from the effective medium theory to an electrical model for granular medium. These authors derived the RGPZ model defined as:

$$k_{RGPZ} = \frac{d^2 \phi^{3m}}{4am^2}$$
(12)

where ϕ is the porosity, m the cementation exponent from the Archie's law, a is a parameter thought to be equal to 8/3 for samples composed of quasi-spherical grains, and d is the relevant grain size. They showed that the relevant grain size is the geometric mean, which can be deduced from Mercury Injection Capillary Pressure (MICP). The relevant grain size can also be inferred from borehole NMR data, and then must be deduced from an empirical procedure relating grain size to the T_2 relaxation time. This new model was shown to match data over 348 samples over a 500 m thick sand-shale succession in the North Sea. Since the porosity can also be derived from NMR data, the advantage of this approach is to provide a log of permeability along the studied borehole, at the scale which is investigated by the NMR tool.

FAULT AND HYDRAULIC FRACTURING

Permeability Prediction within Fault

It has also been proposed to deduce the permeability of the Nojima fault (Japan) using the self-potential observations in surface when water is injected into a well of 1800 m depth (Murakami et al., 2001). Water flow is induced at about 1600 m depth when crossing the fracture zone, and the change in voltage in the aquifer is conducted to the whole part of the well through the iron casing pipe (Fig. 6). Therefore the electrokinetic source occuring at depth can be detected at the surface. Self-potential variations of 10-35 mV in response to water pressure of 35-38x10^5 Pa were observed. The magnitude of self-potential variations decreases with increasing distance from the injection well. An amplitude of −20 mV was detected near the well, about −10 mV at 40 m, and within the noise at one hundred meters. The electrokinetic source is the dragging current expressed by the second term in eq. 2. Assuming the Helmholtz-Smoluchowski equation (eq. 6) and the Darcy's law (second term in eq. 3), and using the definition of the formation factor F, we can write the dragging current:

$$J_{edragg} = -\frac{\epsilon_f \zeta}{Fk} J_f$$

(13)

This dragging current is balanced by the conduction current (the first term in eq. 2). Assuming a line source model with L the length of the casing pipe, the potential difference ΔV between two electrodes at the surface is related to the total conduction current $I_{cond-tot}$ [A] by (Murakami et al., 2001):

$$\Delta V = \frac{I_{cond-tot}}{2\pi\sigma_r L} log(a/b)$$

(14)

where a and b are the distances from the borehole to the electrodes. Then the permeability of the fault is deduced by:

$$k_{fault} = -\frac{\epsilon_f \zeta}{F} \frac{Q_{f-tot}}{I_{cond-tot}}$$

(15)

The total conduction current $I_{cond-tot}$ is deduced from surface potential measurements ΔV (eq. 14). The total water injection (usually several liters/min) provides the value of Q_{f-tot} [m³ s⁻¹]. The formation factor F of the fault can be deduced from resistivity well-logging assuming Archie's law and knowing the fluid conductivity. The value of zeta potential has to be deduced from laboratory experiments published in the literature, possibly using figure 2. Murakami et al. (2001) deduced that the permeability of the fault was higher at the end of the water injection than at the beginning. Assumming different hypotheses for the zeta potential to -1 to -10 mV they deduced a permeability in the range 10^{-16} to 10^{-15} m². The chemical properties of the injected water is important since it can decrease dramatically the zeta potential if species such as Ca^{2+} or Al^{3+} are present in high quantity. The advantage of this method is to be able to deduce the permeability at depth of the fault.

Self-Potentiels Related To Hydraulic Fracturing

Since fluid flow can create streaming potentials, the hydraulic fracturing can induce streaming potentials as the fracture propagates, if the fracture remains fullfilled with water. Laboratory experiments on hydraulic fracturing on granite samples showed that the streaming potential varies linearly with the injection pressure (Moore & Glaser, 2007). However the SPC increases in an exponential trend when approaching the breakdown pressure. Since the permeability also shows an exponential increase with injection pressure, the authors concluded that the SPC is varying as $k^{1.5}$. The explanation was not an effect due to the surface conductivity, but a difference in the hydraulic tortuosity (David, 1993) and electric tortuosity (as suggested by Lorne et al. (1999b)) induced by dilatancy of microcracks. The streaming potential induced by an advancing crack has been modeled by Cuevas et al. (2009). The authors modeled the streaming electric current density by defining a source-time function from the pressure profile in the propagating direction of the opening crack. The streaming electric current is maximum at the tip of the fracture and decays exponentially in front of the tip. The decay constant linearly increases with the propagation speed of the fracture. As the fractures progresses, the streaming potential observed at a distant point results from a superposition of delayed sources arising at the position of the advancing fluid front. The results show that the energy is focused in the vicinity of the advancing fracture's tip, however a tail can also be distinguished as the source behind the tip does not vanish instantaneously. Cuevas et al. (2009) could model the streaming electrical spike recorded by Moore & Glaser (2007) during hydraulic fracturing by modeling the propagation of two cracks and adjusting the propagation velocity, the direction of propagation and the initial fracture volume. The authors concluded that direct information of the hydraulic fracture propagation can be provided by measuring the electrical field at distant.

Hydraulic stimulation is often used to stimulate fluid flow in geothermal reservoirs. Surface electrical potentials were measured when water was injected (during about 7 days) in granite at 5 km depth at the Soultz Hot Dry Rock site (France) (Marquis et al., 2002). An anomalous potential of about 5 mV was interpreted as an electrokinetic effect a depth and measured at the surface because of the conductive well casing. The question of the exact origin between electrokinetic and electrochemical (Maineult, Bernabé & Ackerer, 2006; Maineult, Jouniaux & Bernabé, 2006) effects was raised by Darnet et al. (2004). Finally it has been shown that whatever the injection rate was, the electrochemical contribution was almost negligible (Maineult, Darnet & Marquis, 2006): the SP anomaly was mainly related to the temperature contrast between the in-situ brine and the injected fresh water only at the earliest stage of injection, and was essentially related to water-flows afterwards. Further investigations showed that a slow SP decay is observed after shut-in : its was interpreted as related to large fluid-flow persisting after the end of stimulation and correlated to the microseismic activity (Darnet et al., 2006). The fluid flow was not detected on hydraulic data because it took place in a zone hydraulically disconnected from the openhole. The authors concluded that the SP observations could monitor the fluid flow at the reservoir scale and revealed that the fluid flow plays a major role in the mechanical response of the reservoir to hydraulic stimulation. Another field experiment was performed with periodic pumping tests (injection/production) in a borehole penetrating a sandy aquifer (Maineult et al., 2008). The attenuation of SP amplitude with distance was roughly similar to the pressure attenuation. Therefore the authors proposed that hydraulic diffusivity could be inferred from SP observations. Moreover the comparaison between surface and borehole measurements suggested that nonlinear phenomena are present, probably related to the saturation and desaturation processes occuring in the vadose zone (Maineult et al., 2008).

SEISMO-ELECTROMAGNETIC CONVERSIONS TO DE-TECT HYDRAULIC PROPERTY CONTRASTS

Theoretical Background

The electrokinetic effect can also be induced by seismic wave propagation, which leads to a relative motion between the fluid and the rock matrix. In this case the electrokinetic coefficient depends on the frequency ω as the dynamic permeability $k(\omega)$ (Smeulders et al., 1992). Pride (1994) developed the theory for the coupled electromagnetics and acoustics of porous media. The transport relations [(Pride, 1994) equations (250) and (251)] are:

$$\mathbf{J_e} = \sigma(\omega)\mathbf{E} + \mathcal{L}_{ek}(\omega)\left(-\nabla p + \omega^2 \rho_f \mathbf{u_s}\right)$$

(16)

$$-i\omega\mathbf{J_f} = \mathcal{L}_{ek}(\omega)\mathbf{E} + \frac{k(\omega)}{\eta}\left(-\nabla p + \omega^2 \rho_f \mathbf{u_s}\right)$$

(17)

The electrical fields and mechanical forces which induce the electric current density $\mathbf{J_e}$ and the fluid flow $\mathbf{J_f}$ are, respectively, E and $(-\nabla p + i\omega^2 \rho_f \mathbf{u_s})$, where p is the pore-fluid pressure, us is the solid displacement, E is the electric field, ρ f is the pore-fluid density, and ω is the angular frequency. The electrokinetic coupling $L_{ek}(\omega)$ is now complex and frequency-dependent and describes the coupling between the seismic and electromagnetic fields (Pride, 1994; Reppert et al., 2001):

$$\mathcal{L}_{ek}(\omega) = \mathcal{L}_{ek}\left[1 - i\frac{\omega}{\omega_c}\frac{m}{4}\left(1 - 2\frac{d}{\Lambda}\right)^2\left(1 - i^{3/2}d\sqrt{\frac{\omega\rho_f}{\eta}}\right)^2\right]^{-\frac{1}{2}}$$

(18)

where m and Λ are geometrical parameters of the pores (Λ is defined in Johnson et al. (1987) and m is in the range $4 - 8$), d the Debye length. The transition frequency ω_c defined in the Biot's theory separates the viscous and inertial flow domains and depends on the permeability k_0. The frequency-dependence of the streaming potential coefficient has been studied (Chandler, 1981; Cooke, 1955; Groves & Sears, 1975; Packard, 1953; Reppert et al., 2001; Schoemaker et al., 2007; 2008; Sears & Groves, 1978) mainly on synthetic samples. Both models (Gao & Hu, 2010; Garambois & Dietrich, 2001; 2002; Haartsen et al., 1998; Haartsen & Pride, 1997; Pain et al., 2005; Schakel &

Smeulders, 2010) and laboratory experiments (Block & Harris, 2006; Bordes et al., 2006; 2008; Chen & Mu, 2005; Zhu et al., 1999) have been developed on these seismoelectromagnetic conversions. Note that assuming the Helmholtz-Smoluchowski equation for the streaming potential coefficient leads to the electrokinetic coupling inversely dependent on the formation factor F as:

$$\mathcal{L}_{ek} = \frac{\epsilon_f \zeta}{\eta_f F}$$

(19)

The formation factor is inversely related to the permeability and proportionnal to the hydraulic radius $F = CR^2/k_0$ (Paterson, 1983). Since the permeability can vary of about fifteen orders of magnitude, whereas this is not the case of the hydraulic radius, the static electrokinetic coupling L_{ek} will increase with increasing permeability

Detection of Permeability Contrasts

Two kinds of mechanical to electromagnetic conversions exist: 1) The electrokinetic signal which travels with the acoustic wave; 2) The interfacial conversion occuring at contrasts of physical properties such as permeability.

The first kind of conversion has been used to show that a reliable permeability log can be derived from electrokinetic measurements (Singer et al., 2005), using an acoustic source within a borehole (Fig. 7). Singer et al. (2005) showed by a finite element model and by laboratory experiments that the normalized coefficient defined by the electric field divided by the pressure depends [V Pa^{-1} m^{-1}] on the permeability. This coefficient is coherent with the electrokinetic coefficient L_{ek} (eq. 19) per unit of conductance [S] and then should increase with increasing permeability. At low permeability the oscillating source will induce a larger solid displacement because the fluid is not easily displaced. However the relative movement between solid and fluid is limited, leading to a decrease of the electric field even if pressure increases, so that this normalized coefficient is decreased. The investigated depth of such a permeability is of the order of centimeters. The source was a short steel tube near the top of the borehole and hit on top with a hammer. The main wave propagation is a Stoneley wave which induces the electric field. The logging tool is moved step-by-step within the borehole (Fig. 7). This model showed that the normalized coefficient could detect a 0.5 m-thick bed of permeability 10^{-13}m^2 within a formation of permeability 10^{-15} m^2 . The measured amplitude of the normalized coefficient on sandstones is in the range 1.6×10^{-7} to 2.5×10^{-6} [V Pa^{-1} m^{-1}] increasing with increasing permeabilities from $6.2 \ 10^{-15}$ m^2 to 2.2×10^{-12} m^2 .

The second kind of conversion can be used to detect contrasts in permeability in the crust. The seismic source induces a seismic wave propagation downward up to the interface (Fig. 8). Because of the difference in the physical properties there is a charge inbalance that causes a charge separation on both sides of the interface. This acts as en electric dipole which emits an electromagnetic wave that travels with the speed of the light in the medium and that can be detected at the surface (Fig. 9). The velocity of the seismic wave propagation is deduced by surface measurements of the soil velocity. Then the depth of the interface can be deduced by picking the time arrival of the electromagnetic wave. Usually the seismoelectric signals show low amplitude from 100 μV to mV. Then signal processing needs filtering techniques such as Butler & Russell (1993). The advantage of this method is to detect the contrasts of permeability at depth from few meters to few hundreds of meters (Dupuis & Butler, 2006; Dupuis et al., 2007; Dupuis et al., 2009; Haines, Guitton & Biondi, 2007; Haines, Pride, Klemperer & Biondi, 2007; Strahser et al., 2007; 2011; Thompson et al., 2005).

LIMITATIONS OF THIS TECHNIQUE

The limitations of this technique arise from the low amplitude of the electrical signal. It needs good pre-amplifiers to be able to detect the signals. Then it needs an adapted signal processing to remove the anthropic noise, and further filtering techniques to extract the expected signal from the remaining records. The interpretation of self-potential observations may not be easy if the signals are induced not only by the electrokinetic effect, but also by a thermoelectric effect, and by an electrochemical effect. The interpretation of the seismo-electric conversion may not be easy if the contrast in the permeability is not high enough.

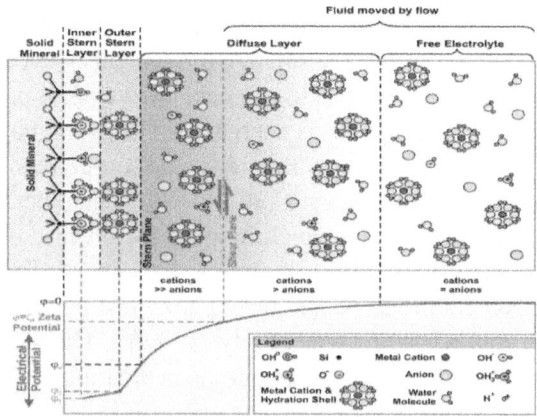

Figure 1. Electric double layer, courtesy of P.W.J. Glover (Glover & Jackson, 2010). The solid mineral presented is the case of silica. At pH above the iso-

electric point the cations are adsorbed within the Stern layer; there is an excess of cations in the diffuse layer. The zeta potential is defined at the shear plane. The fluid flow creates a streaming current which is balanced by the conduction current, leading to the streaming potential.

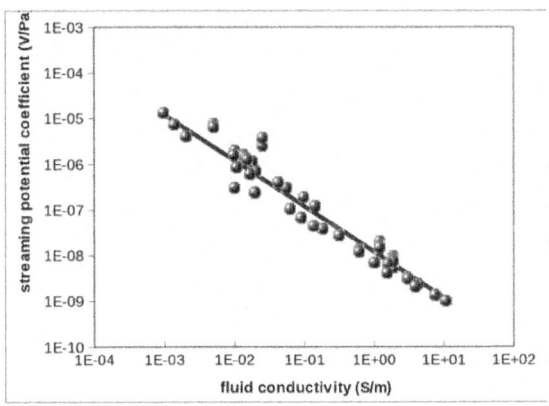

Figure 2. Streaming potential coefficient from data collected (in absolute value) on sands and sandstones at pH 7-8 (when available) from Ahmad (1964); Guichet et al. (2006; 2003); Ishido & Mizutani (1981); Jaafar et al. (2009); Jouniaux & Pozzi (1997); Li et al. (1995); Lorne et al. (1999a); Pengra et al. (1999); Perrier & Froidefond (2003). The regression (black line) leads to $C_{s0} = -1.2 \times 10^{-8} \sigma^{-1}_f$. A zeta potential of -17 mV can be inferred from these collected data (from Allègre et al. (2010)).

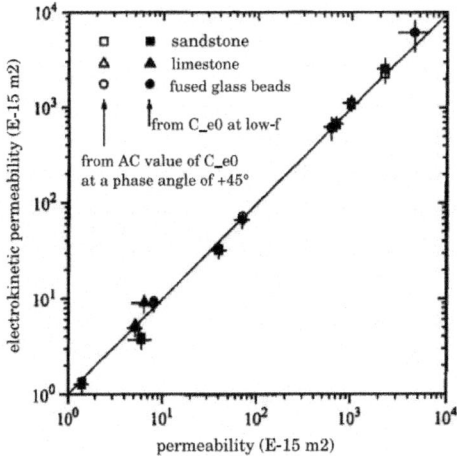

Figure 3. Comparison between the permeability k and the electrokinetic permeability k_e. The solid line is $k_e = k$ (modified from Pengra et al. (1999)).

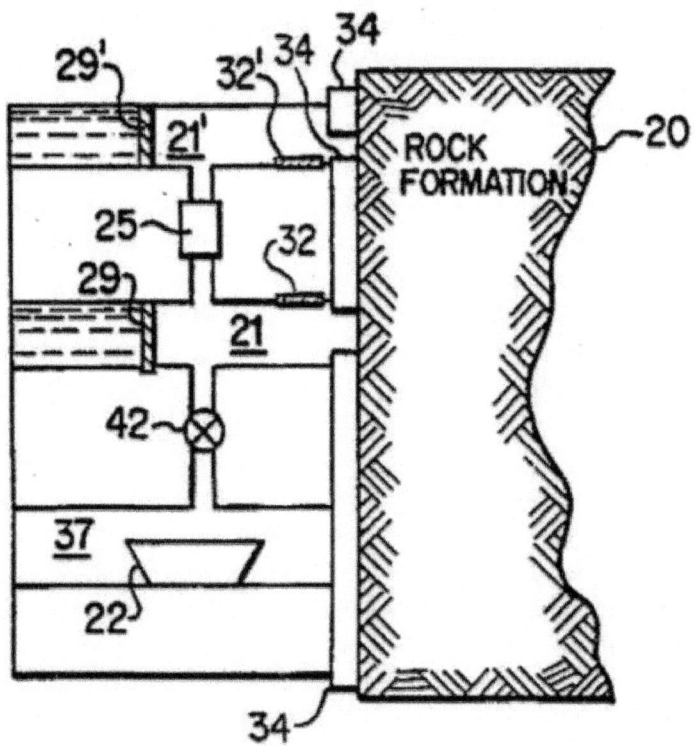

Figure 4. In-situ permeability measurement (from Wong (1995)) from streaming potential and electro-osmosis measurements using eq. 11. For the streaming potential measurement: an oscillating presure is applied by electromechanical transducer (22) to the rock formation (20) through fluid chamber (21) with valve (42) open. The pressure differential in the rock between fluid chamber (21) and (21') is measured by a pressure sensor (25) and the induced voltage difference is measured by the voltage electrodes (32) and (32'). For the electro-osmosis measurements the valve (42) is closed, the pressure difference induced when a current is passed through the rock (by current electrodes 29 and 29') is measured by pressure sensor (25).

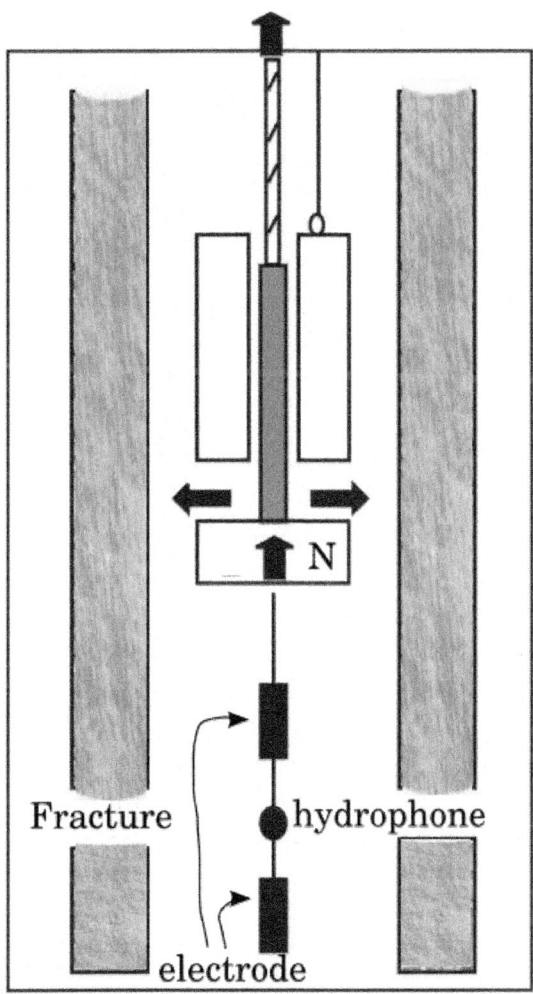

Figure 5. Scheme of the principle of borehole electrokinetic response to detect fractures (modified from Hunt & Worthington (2000)). The source is a nylon block (N) pulled by the rope, which induces fluid flow near the wall of the rock formation, leading to an electrokinetic effect. The electrodes are 1 m apart. The hydrophone is 2.5 m below the source. The electrode response is normalized by the peak pressure.

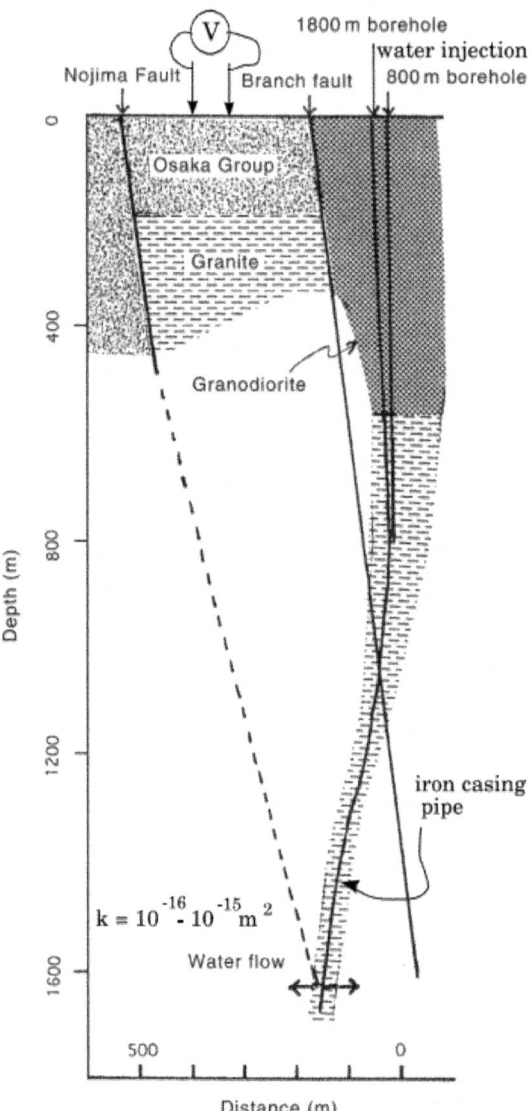

Figure 6. Measurement of the permeability of the Nojima fault (modified from Murakami et al. (2001)). The water injection inside the borehole of 1800 m depth crosses the fault inducing an electrokinetic source at depth within the fault. The conduction current is conducted by the iron pipe up to the surface. The difference of potential V is measured by electrodes on the surface. The permeability is deduced from eq. 15

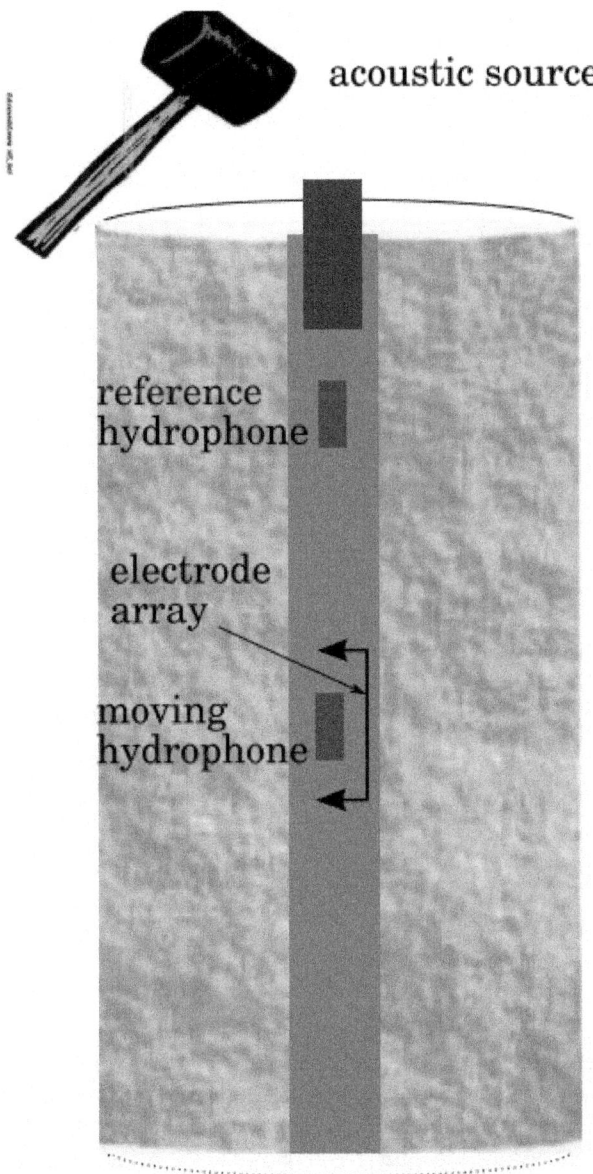

Figure 7. Scheme of the principle of electrokinetic logging to measure the permeability (modified from Singer et al. (2005)). The acoustic source induces a Stoneley wave propagation (detected by the hydrophones) leading to an electric field (measured by the electrodes). The experiment is repeated by moving the tool downward.

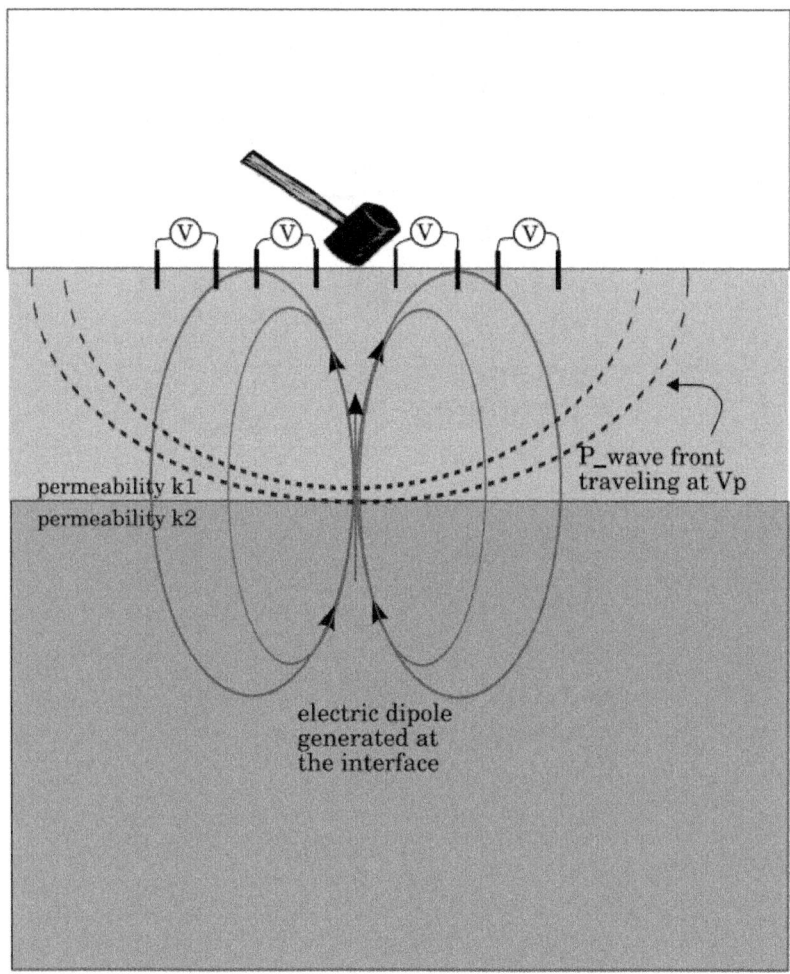

permeability k1

permeability k2

P_wave front
traveling at Vp

electric dipole
generated at
the interface

Figure 8. The seismic waves propagates up to the interface where an electric dipole is generated because of the contrast in permeability. This electromagnetic wave can be detected at the surface by measuring the difference of the electrical potential V between electrodes. Picking the time arrival allows to know the depth of the interface.

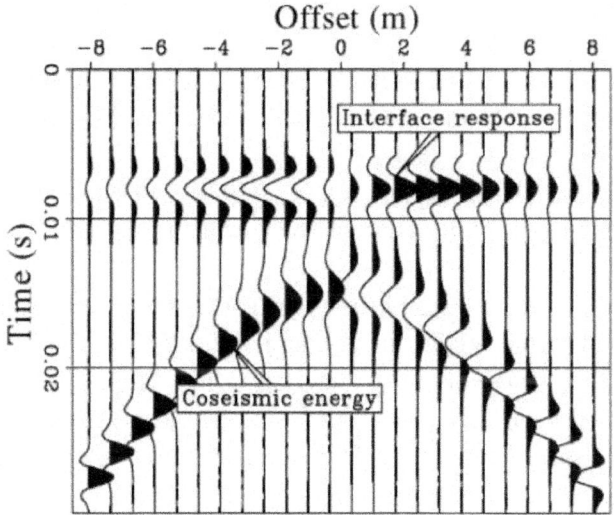

Figure 9. Model of the seimoelectric response to a hammer strike on the surface at position zero (from Haines (2004)). The seismoelectric signal is shown as measured at the surface along a line centered on the seismic source. The interfacial signal is related to a contrast between properties of the media, such as the permeability.

CONCLUSION

The electrokinetic properties can be used to deduce permeability in the crust, possibly at depth, within fault, and along boreholes. Some conditions are needed to be able to use electrokinetic coupling to infer hydraulic properties. The electrical noise can prevent being able to detect small electric potentials, even using appropriate filtering techniques. When possible, the joint inversion with other observations can improve parameters such as electrical conductivity. The seismoelectric method could provide deeper investigations when using stronger seismic sources.

REFERENCES

1. Adamson, A. W. (1976). Physical chemistry of surfaces, John Wiley and sons, New York.

2. Allègre, V., Jouniaux, L., Lehmann, F. & Sailhac, P. (2010). Streaming Potential dependence on water-content in fontainebleau sand, Geophys. J. Int. 182: 1248–1266.

3. Allègre, V., Jouniaux, L., Lehmann, F. & Sailhac, P. (2011). Reply to the comment by A. Revil and N. Linde on: "Streaming potential dependence

on water-content in fontainebleau sand" by Allègre et al., Geophys. J. Int. 186: 115–117.

4. Archie, G. E. (1942). The electrical resistivity log as an aid in determining some reservoir characteristics, Trans. Am. Inst. Min. Metall. Pet. Eng. (146): 54–62.

5. Aubert, M. & Atangana, Q. Y. (1996). Self-potential method in hydrogeological exploration of volcanic areas, Ground Water 34: 1010–1016.

6. Auriault, J. & Strzelecki, T. (1981). On the electro-osmotic flow in saturated porous media, Int. J. Engrg. Sci. 19: 915–928.

7. Beddiar, K., Berthaud, Y. & Dupas, A. (2002). Experimental verification of the onsager's reciprocal relations for electro-osmosis and electro-filtration phenomena on a saturated clay, C. R. Mécanique 330: 893–898.

8. Beddiar, K., Fen-Chong, T., Dupas, A., Berthaud, Y. & Dangla, P. (2005). Role of ph in electro-osmosis: Experimental study on nacl-water saturated kaolinite, Transport in Porous media 61: 93–107.

9. Block, G. I. & Harris, J. G. (2006). Conductivity dependence of seismoelectric wave phenomena in fluid-saturated sediments, J. Geophys. Res. 111: B01304.

10. Bordes, C., Jouniaux, L., Dietrich, M., Pozzi, J.-P. & Garambois, S. (2006). First laboratory measurements of seismo-magnetic conversions in fluid-filled Fontainebleau sand, Geophys. Res. Lett. 33: L01302.

11. Bordes, C., Jouniaux, L., Garambois, S., Dietrich, M., Pozzi, J.-P. & Gaffet, S. (2008). Evidence of the theoretically predicted seismo-magnetic conversion, Geophys. J. Int. 174: 489–504.

12. Butler, K. E. & Russell, R. D. (1993). Substraction of powerline harmonics from geophysical records, Geophysics 58: 898–903.

13. Chandler, R. (1981). Transient streaming potential measurements on fluid-saturated porous structures: An experimental verification of Biot's slow wave in the quasi-static limit, J. Acoust. Soc. Am. 70: 116–121.

14. Chen, B. & Mu, Y. (2005). Experimental studies of seismoelectric effects in fluid-saturated porous media, J. Geophys. Eng. 2: 222–230.

15. Cooke, C. E. (1955). Study of electrokinetic effects using sinusoidal pressure and voltage, J. Chem. Phys. (23): 2299–2303.

16. Cuevas, N., Moore, J. & Glaser, S. (2009). Electrokinetic coupling in hydraulic fracture propagation, SEG Technical Program Expanded Abstracts 28: 1721–1725.

17. Darnet, M., G.Marquis & Sailhac, P. (2006). Hydraulic stimulation of geothermal reservoirs:fluid flow, electric potential and microseismicity relationships, Geophys. J. Int. 166: 438–444.

18. Darnet, M., Maineult, A. & Marquis, G. (2004). On the origins of self-potential (sp) anomalies induced by water injections into geothermal reservoirs, Geophys. Res. Lett. 31: L19609.

19. Darnet, M. & Marquis, G. (2004). Modelling streaming potential (sp) signals induced by water movement in the vadose zone, J. Hydrol. 285: 114–124.

20. David, C. (1993). Geometry of flow paths for fluid transport in rocks, J. Geophys. Res. 98: 12267–12278.

21. Davis, J. A., James, R. O. & Leckie, J. (1978). Surface ionization and complexation at the oxide/water interface, J. Colloid Interface Sci. 63: 480–499.

22. Davis, J. & Kent, D. (1990). Surface complexation modeling in aqueous geochemistry, in Mineral Water Interface Geochemistry, M.F. Hochella and A.F. White, Mineralogical Society of America.

23. Doussan, C. & Ruy, S. (2009). Prediction of unsaturated soil hydraulic conductivity with electrical conductivity, Water Resources Res. 45: W10408.

24. Dukhin, S. S. & Derjaguin, B. V. (1974). Surface and Colloid Science, edited by E. Matijevic, John Wiley and sons, New York.

25. Dupuis, J. C. & Butler, K. E. (2006). Vertical seismoelectric profiling in a borehole penetrating glaciofluvial sediments, Geophys. Res. Lett. 33.

26. Dupuis, J. C., Butler, K. E. & Kepic, A. W. (2007). Seismoelectric imaging of the vadose zone of a sand aquifer, Geophysics 72: A81–A85.

27. Dupuis, J. C., Butler, K. E., Kepic, A. W. & Harris, B. D. (2009). Anatomy of a seismoelectric conversion: Measurements and conceptual modeling in boreholes penetrating a sandy aquifer, J. Geophys. Res. Solid Earth 114(B13): B10306.

28. El-kaliouby, H. & Al-Garni, M. (2009). Inversion of self-potential anomalies caused by 2d inclined sheets using neural networks, J. Geophys. Eng. 6: 29–34.

29. Fagerlund, F. & Heison, G. (2003). Detecting subsurface grounwater flow in fractured rock using self-potential (sp) methods, Environmental Geology 43: 782–794.

30. Fernandez-Martinez, J., Garcia-Gonzalo, E. & Naudet, V. (2010). Particle swarm optimization applied to solving and appraising the streaming potential inverse problem, Geophysics 75: WA3–WA15.

31. Finizola, A., Sortino, F., L/'enat, J.-F. & Aubert, M. (2003). The summit hydrothermal system of stromboli, new insights from self-potential temperature, c02 and fumarolic fluid measurements, with structural and monitoring implications, Bulletin of Volcanology (65): 486–504.

32. Gao, Y. & Hu, H. (2010). Seismoelectromagnetic waves radiated by a double couple source in a saturated porous medium, Geophys. J. Int. 181: 873–896.

33. Garambois, S. & Dietrich, M. (2001). Seismoelectric wave conversions in porous media: Field measurements and transfer function analysis, Geophysics 66: 1417–1430.

34. Garambois, S. & Dietrich, M. (2002). Full waveform numerical simulations of seismoelectromagnetic wave conversions in fluid-saturated stratified porous media, J. Geophys. Res. 107(B7): ESE 5–1.

35. Gibert, D. & Pessel, M. (2001). Identification of sources of potential fields with the continuous wavelet transform: Application to self-potential profiles, Geophys. Res. Lett. 28: 1863–1866.

36. Gibert, D. & Sailhac, P. (2008). Comment on: Self-potential signals associated with preferential grounwater flow pathways in sinkholes, by A. Jardani J.P dupont A. Revil, J. Geophys. Res. 113: B03210.

37. Glover, P. & Jackson, M. (2010). Borehole electrokinetics, The Leading Edge pp. 724–728.

38. Glover, P. W. J. & Walker, E. (2009). Grain-size to effective pore-size transformation derived from electrokinetic theory, Geophysics 74: E17–E29.

39. Glover, P. W. J., Zadjali, I. I. & Frew, K. A. (2006). Permeability prediction from MICP and NMR data using an electrokinetic approach, Geophysics 71: F49–F60.

40. Groves, J. & Sears, A. (1975). Alternating streaming current measurements, J. Colloid Interface Sci. 53: 83–89.

41. Guichet, X., Jouniaux, L. & Catel, N. (2006). Modification of streaming potential by precipitation of calcite in a sand-water system: laboratory measurements in the pH range from 4 to 12, Geophys. J. Int. 166: 445–460.

42. Guichet, X., Jouniaux, L. & Pozzi, J.-P. (2003). Streaming potential of a sand column in partial saturation conditions, J. Geophys. Res. 108(B3): 2141.

43. Haartsen, M. W., Dong, W. & Toksöz, M. N. (1998). Dynamic streaming currents from seismic point sources in homogeneous poroelastic media, Geophys. J. Int. 132: 256–274.

44. Haartsen, M. W. & Pride, S. (1997). Electroseismic waves from point sources in layered media, J. Geophys. Res. 102: 24,745–24,769.

45. Haines, S. (2004). Seismoelectric imaging of shallow targets, PhD dissertation (Stanford University).

46. Haines, S. S., Guitton, A. & Biondi, B. (2007). Seismoelectric data processing for surface surveys of shallow targets, Geophysics 72: G1–G8.

47. Haines, S. S., Pride, S. R., Klemperer, S. L. & Biondi, B. (2007). Seismoelectric imaging of shallow targets, Geophysics 72: G9–G20.

48. Henry, P., Jouniaux, L., Screaton, E. J., S.Hunze & Saffer, D. M. (2003). Anisotropy of electrical conductivity record of initial strain at the toe of the Nankai accretionary wedge, J. Geophys. Res. 108: 2407.

49. Hunt, C. W. & Worthington, M. H. (2000). Borehole elektrokinetic responses in fracture dominated hydraulically conductive zones, Geophys. Res. Lett. 27(9): 1315–1318.

50. Hunter, R. (1981). Zeta Potential in Colloid Science: Principles and Applications, Academic., New York.

51. Ishido, T. & Mizutani, H. (1981). Experimental and theoretical basis of electrokinetic phenomena in rock water systems and its applications to geophysics, J. Geophys. Res. 86: 1763–1775.

52. Ishido, T., Mizutani, H. & Baba, K. (1983). Streaming potential observations, using geothermal wells and in situ electrokinetic coupling coefficients under high temperature, Tectonophysics 91: 89–104.

53. Ishido, T. & Pritchett, J. (1999). Numerical simulation of electrokinetic potentials associated with subsurface fluid flow, J. Geophys. Res. 104(B7): 15247–15259.

54. Jaafar, M. Z., Vinogradov, J. & Jackson, M. D. (2009). Measurement of streaming potential coupling coefficient in sandstones saturated with high salinity nacl brine, Geophys. Res. Lett. 36.

55. Jackson, M. D. (2010). Multiphase electrokinetic coupling: Insights into the impact of fluid and charge distribution at the pore scale from a bundle of capillary tubes model, J. Geophys. Res. 115: B07206.

56. Johnson, D. L., Koplik, J. & Dashen, R. (1987). Theory of dynamic permeability in fluid saturated porous media, J. Fluid. Mech. 176: 379–402.

57. Jouniaux, L., Bernard, M.-L., Zamora, M. & Pozzi, J.-P. (2000). Streaming potential in volcanic rocks from Mount Peleé, J. Geophys. Res. 105: 8391–8401.

58. Jouniaux, L., Lallemant, S. & Pozzi, J. (1994). Changes in the permeability, streaming potential and resistivity of a claystone from the Nankai prism under stress, Geophys. Res. Lett. 21: 149–152.

59. Jouniaux, L., Maineult, A., Naudet, V., Pessel, M. & Sailhac, P. (2009). Review of self-potential methods in hydrogeophysics, C.R. Geosci. 341: 928–936.

60. Jouniaux, L. & Pozzi, J.-P. (1995a). Permeability dependence of streaming potential in rocks for various fluid conductivity, Geophys. Res. Lett. 22: 485–488.

61. Jouniaux, L. & Pozzi, J.-P. (1995b). Streaming potential and permeability of saturated sandstones under triaxial stress: consequences for electrotelluric anomalies prior to earthquakes, J. Geophys. Res. 100: 10,197–10,209.

62. Jouniaux, L. & Pozzi, J.-P. (1997). Laboratory measurements anomalous 0.1-0.5 Hz streaming potential under geochemical changes: Implications for electrotelluric precursors to earthquakes, J. Geophys. Res. 102: 15,335–15,343.

63. Jouniaux, L., Pozzi, J.-P., Berthier, J. & Massé, P. (1999). Detection of fluid flow variations at the Nankai trough by electric and magnetic measurements in boreholes or at the seafloor, J. Geophys. Res. 104: 29293–29309.

64. Jouniaux, L., Zamora, M. & Reuschlé, T. (2006). Electrical conductivity evolution of non-saturated carbonate rocks during deformation up to failure, Geophys. J. Int. 167: 1017–1026.

65. Lesmes, D. P. & Friedman, S. P. (2005). Relationships between the electrical and hydrogeological properties of rocks and soils, Hydrogeophysics, Springer, Dordrecht, The Netherlands, chapter 4, pp. 87–128.

66. Li, S., Pengra, D. & Wong, P. (1995). Onsager's reciprocal relation and the hydraulic permeability of porous media, Physical Review E 51(6): 5748–5751.

67. Lorne, B., Perrier, F. & Avouac, J.-P. (1999a). Streaming potential measurements. 1. properties of the electrical double layer from crushed rock samples, J. Geophys. Res. 104(B8): 17,857–17,877.

68. Lorne, B., Perrier, F. & Avouac, J.-P. (1999b). Streaming potential measurements. 2. relationship between electrical and hydraulic flow patterns from rocks samples during deformations, J. Geophys. Res. 104(B8): 17,879–17,896.

69. Maineult, A., Bernabé, Y. & Ackerer, P. (2006). Detection of advected, recating redox fronts from self-potential measurements, J. Contaminant Hydrology (86): 32–52.

70. Maineult, A., Darnet, M. & Marquis, G. (2006). Correction to on the origins of self-potential (sp) anomalies induced by water injections into geothermal reservoirs, Geophys. Res. Lett. (33): L20319.

71. Maineult, A., Jouniaux, L. & Bernabé, Y. (2006). Influence of the mineralogical composition on the self-potential response to advection of kcl concentration fronts through sand, Geophys. Res. Lett. (33): L24311.

72. Maineult, A., Strobach, E. & Renner, J. (2008). Self-potential signals induced by periodic pumping, J. Geophys. Res. 113: B01203.

73. Marquis, G., Darnet, M., Sailhac, P., Singh, A. K. & Gérard, A. (2002). Surface electric variations induced by deep hydraulic stimulation: an example from the soultz hdr site, Geophys. Res. Lett. 29.

74. Miller, D. (1960). Thermodynamics of irreversible processes, the experimental verification of onsager reciprocal relations, Chem. Rev. 60(1): 15–37.

75. Minsley, B., Sogade, J. & Morgan, F. (2007). Three-dimensional modelling source inversion of self-potential data, J. Geophys. Res. 112: B02202.

76. Moore, J. & Glaser, S. (2007). Self-potential observations during hydraulic fracturing, J. Geophys. Res. 112: B02204.

77. Moreau, F., Gibert, D. & Saracco, G. (1996). Filtering non-stationnary geophysical data with orthogonal wavelets, Geophys. Res. Lett. 23(4): 407–410.

78. Murakami, H., Hashimoto, T., N.Oshiman, Yamaguchi, S., Honkuba, Y. & Sumitomo, N. (2001). Electrokinetic phenomena associated with a water injection experiment at the nojima fault on awaji island, japan, The Island Arc 10: 244–251.

79. Naudet, V., Fernandez-Martinez, J., Garcia-Gonzalo, E. & Fernandez-Alvarez, J. (2008). Estimation of water table from self-potential data using particle swarm optimization (pso), SEG Expanded Abstracts 27: 1203.

80. Onsager, L. (1931). Reciprocal relations in irreversible processes:i, Phys. Rev. 37: 405–426.

81. Overbeek, J. T. G. (1952). Electrochemistry of the double layer, Colloid Science, Irreversible Systems, edited by H. R. Kruyt, Elsevier 1: 115–193.

82. Packard, R. G. (1953). Streaming potentials across capillaries for sinusoidal pressure, J. Chem. Phys 1(21): 303–307.

83. Pain, C., Saunders, J. H., Worthington, M. H., Singer, J. M., Stuart-Bruges, C. W., Mason, G. & Goddard., A. (2005). A mixed finite-element method for solving the poroelastic Biot equations with electrokinetic coupling, Geophys. J. Int. 160: 592–608.

84. Paterson, M. (1983). The equivalent channel model for permeability and resistivity in fluid-saturated rock- a re-appraisal, Mechanics of Materials 2: 345–352.

85. Pengra, D. B., Li, S. X. & Wong, P.-Z. (1999). Determination of rock properties by low frequency ac electrokinetics, J. Geophys. Res. 104(B12): 29.485–29.508.

86. Perrier, F. E., Petiau, G., Clerc, G., Bogorodsky, V., Erkul, E., Jouniaux, L., Lesmes, D., Magnae, J., Meunier, J.-M., Morgan, D., Nascimento, D., Oettinger, G., Schwartz, G., Toh, H., Valiant, M.-J., Vozoff, K. & Yazici-Cakin, O. (1997). A one-year systematic study of electrodes for long period measurements of the electric field in geophysical environments, J. Geomag. Geoelectr 49: 1677–1696.

87. Perrier, F. & Froidefond, T. (2003). Electrical conductivity and streaming potential coefficient in a moderately alkaline lava series, Earth and Planetary Science Letters 210: 351–363.

88. Perrier, F. & Morat, P. (2000). Characterization of electrical daily variations induced by capillary flow in the non-saturated zone, Pure Appl. Geophys. 157: 785–810.

89. Perrier, F., Trique, M., Lorne, B., Avouac, J.-P., Hautot, S. & Tarits, P. (1998). Electric potential variations associated with lake variations, Geophys. Res. Lett. 25: 1955–1958.

90. Petiau, G. (2000). Second generation of lead-lead chloride electrodes for geophysical applications, Pure Appl. Geophys. 3: 357–382.

91. Petiau, G. & Dupis, A. (1980). Noise, temperature coefficient and long time stability of electrodes for telluric observations, Geophys. Prospect. 28(5): 792–804.

92. Pezard, P., Gautier, S., Borgne, T. L., Legros, B. & Deltombe, J.-L. (2009). Muset: A multiparameter and high precision sensor for downhole spontaneous electrical potential measurements, Comptes Rendus - Geoscience 341: 957–964.

93. Pinettes, P., Bernard, P., Cornet, F., Hovhannissian, G., Jouniaux, L., Pozzi, J.-P. & Barthés, V. (2002). On the difficulty of detecting streaming potentials generated at depth, Pure Appl. Geophys. 159: 2629–2657.

94. Pride, S. (1994). Governing equations for the coupled electromagnetics and acoustics of porous media, Phys. Rev. B: Condens. Matter 50: 15678–15695.

95. Pride, S. & Morgan, F. D. (1991). Electrokinetic dissipation induced by seismic waves, Geophysics 56(7): 914–925.

96. Reppert, P. M., Morgan, F. D., Lesmes, D. P. & Jouniaux, L. (2001). Frequency-dependent streaming potentials, J. Colloid Interface Sci. (234): 194–203.

97. Sailhac, P., Darnet, M. & Marquis, G. (2004). Electrical streaming potential measured at the ground surface: forward modeling and inversion issues for monitoring infiltration and characterizing the vadose zone, Vadose Zone J. (3): 1200–1206.

98. Saracco, G., Labazuy, P. & Moreau, F. (2004). Localization of self-potential sources in volcano-electric effect with complex continuous wavelet transform and electrical tomography methods for an active volcano, Geophys. Res. Lett. (31): L12610.

99. Saunders, J. H., Jackson, M. D. & Pain, C. C. (2008). Fluid flow monitoring in oilfields using downhole measurements of electrokinetic potential, Geophysics 73: E165–E180.

100. Schakel, M. & Smeulders, D. (2010). Seismoelectric reflection and transmission at a fluid/porous-medium interface, J. Acoust. Soc. Am. 127: 13–21.

101. Schoemaker, F., Smeulders, D. & Slob, E. (2007). Simultaneous determination of dynamic permeability and streaming potential, SEG expanded abstracts 26: 1555–1559.

102. Thompson, A., Hornbostel, S., Burns, J., Murray, T., Raschke, R., Wride, J., McCammon, P., Sumner, J., Haake, G., Bixby, M., Ross, W., White, B., Zhou, M. & Peczak, P. (2005). Field tests of electroseismic hydrocarbon detection, SEG Technical Program Expanded Abstracts .

103. Vinogradov, J., Jaafar, M. & Jackson, M. D. (2010). Measurement of streaming potential coupling coefficient in sandstones saturated with natural and artificial brines at high salinity, J. Geophys. Res. 115: B12204.

104. Wong, P. (1995). Determination of permeability of porous media by streaming potential and electro-osmotic coefficients, United States Patent Number 5,417,104.

105. Zhu, Z., Haartsen, M. W. & Toksöz, M. N. (1999). Experimental studies of electrokinetic conversions in fluid-saturated borehole models, Geophysics 64: 1349–1356.

CITATION

CHAPTER 1

Yongfu Xu (2013). Unsaturated Hydraulic Conductivity of Fractal-Textured Soils, Hydraulic Conductivity, Dr. Vanderlei Rodrigues Da Silva (Ed.), ISBN: 978-953-51-1208-2, InTech, DOI: 10.5772/56716.

CHAPTER 2

Peng Wu, Jueyi Sui and Ram Balachandar (2015). The Impact of Ice Cover and Sediment Nonuniformity on Erosion Around Hydraulic Structures, Effects of Sediment Transport on Hydraulic Structures, Prof. Vlassios Hrissanthou (Ed.), ISBN: 978-953-51-2231-9, InTech, DOI: 10.5772/61468.

CHAPTER 3

Marie-France Jutras and Paul A. Arp (2013). Role of Hydraulic Conductivity Uncertainties in Modeling Water Flow through Forest Watersheds, Hydraulic Conductivity, Dr. Vanderlei Rodrigues Da Silva (Ed.), ISBN: 978-953-51-1208-2, InTech, DOI: 10.5772/56900.

CHAPTER 4

Vincent M.C. (2013). Five Things You Didn't Want to Know about Hydraulic Fractures, Effective and Sustainable Hydraulic Fracturing, Dr. Rob Jeffrey (Ed.), ISBN: 978-953-51-1137-5, InTech, DOI: 10.5772/56066.

CHAPTER 5

Sebastian Brenne, Michael Molenda, Ferdinand Stöckhert and Michael Alber (2013). Hydraulic and Sleeve Fracturing Laboratory Experiments on 6 Rock Types, Effective and Sustainable Hydraulic Fracturing, Dr. Rob Jeffrey (Ed.), ISBN: 978-953-51-1137-5, InTech, DOI: 10.5772/56301.

CHAPTER 6

Charles Fairhurst (2013). Fractures and Fracturing - Hydraulic fracturing in Jointed Rock, Effective and Sustainable Hydraulic Fracturing, Dr. Rob Jeffrey (Ed.), ISBN: 978-953-51-1137-5, InTech, DOI: 10.5772/56366.

CHAPTER 7

Cheng-Yu Ku and Shih-Meng Hsu (2011). Estimating Hydraulic Conductivity of Highly Disturbed Clastic Rocks in Taiwan, Hydraulic Conductivity - Issues, Determination and Applications, Prof. Lakshmanan Elango (Ed.), ISBN: 978-953-307-288-3, InTech, DOI: 10.5772/18553.

CHAPTER 8

Levent Yilmaz (2015). General Hydraulic Geometry, Effects of Sediment Transport on Hydraulic Structures, Prof. Vlassios Hrissanthou (Ed.), ISBN: 978-953-51-2231-9, InTech, DOI: 10.5772/61643.

CHAPTER 9

Olga Kresse and Xiaowei Weng (2013). Hydraulic Fracturing in Formations with Permeable Natural Fractures, Effective and Sustainable Hydraulic Fracturing, Dr. Rob Jeffrey (Ed.), ISBN: 978-953-51-1137-5, InTech, DOI: 10.5772/56446.

CHAPTER 10

Oagile Dikinya (2011). Hydraulic Conductivity of Semi-Quasi Stable Soils: Effects of Particulate Mobility, Developments in Hydraulic Conductivity Research, Dr. Oagile Dikinya (Ed.), ISBN: 978-953-307-470-2, InTech, DOI: 10.5772/15681.

CHAPTER 11

Laurence Jouniaux (2011). Electrokinetic Techniques for the Determination of Hydraulic Conductivity, Hydraulic Conductivity - Issues, Determination and Applications, Prof. Lakshmanan Elango (Ed.), ISBN: 978-953-307-288-3, InTech, DOI: 10.5772/17599.

INDEX

A

Acoustic Emission (AE) 134
ADV (Acoustic Doppler Velocimetry) 72

D

Discrete fracture networks (DFN's) 151

E

electrokinetic 251, 252, 253, 255, 256, 257, 259, 260, 261, 262, 263, 265, 266, 267, 269, 270, 272, 273, 276, 277, 278
Enhanced Geothermal Energy (EGS) 144
Enhanced Geothermal Systems (EGS 142
exchangeable sodium percentage (ESP) 237

G

gouge content designation (GCD) 180, 191
Gouge Content Designation (GCD) 192

H

homogeneous 115, 117
Hydraulic condtivity (HC) 240
hydraulic conductivity (HC) 245
hydraulic fracture (HF) 214
Hydraulic Fracturing (HF) 148
hydrocarbons 115, 116, 117, 122, 123
hydrological models 1
hydrology 199

L

Linearly Elastic Fracture Mechanics (LEFM) 146
lithology permeability index (LPI) 180, 191

M

Meyer-Peter and Muller (MPM) 204

N

natural fracture (NF) 214

O

organic matter (OM) 94

P

Particle Flow Code (PFC) code) 151
pore-size distribution (PSD) 2, 7, 8, 10,
     14, 15, 19, 20, 21, 27, 28, 29, 30,
     41

R

relative hydraulic conductivity (RHC) 2,
     3, 26, 27, 29, 30, 31, 35, 37, 38,
     39, 41, 42, 247
rock mass 179, 180, 184, 185, 191, 192,
     193, 194, 196, 197
rock quality designation (RQD) 180,
     191

S

signal-to-noise ratio (SNR) 75
sodium adsorption ratio (SAR) 245, 247
Sodium adsorption ratio (SAR) 238
Soil hydraulic 93
soil matrix 238, 246
soil-water characteristic curve (SWCC)
     1, 2, 17, 18, 19, 21, 24, 28, 29, 30,
     31, 35, 37, 39, 42, 44
soil-water diffusivity (SWD) 1, 27, 29,
     30

T

turbulent kinetic energy (TKE) 86

U

Unconventional Fracture Model (UFM*)
     213